Lecture Notes in Physics

For information about Vols. 1–90, please contact your bookseller or Springer-Verlag.

Vol. 91: Computing Methods in Applied Sciences and Engineering, 1977, II. Proceedings, 1977. Edited by R. Glowinski and J. L. Lions. VI, 359 pages. 1979.

Vol. 92: Nuclear Interactions. Proceedings, 1978. Edited by B. A. Robson. XXIV, 507 pages. 1979.

Vol. 93: Stochastic Behavior in Classical and Quantum Hamiltonian Systems. Proceedings, 1977. Edited by G. Casati and J. Ford. VI, 375 pages. 1979.

Vol. 94: Group Theoretical Methods in Physics. Proceedings, 1978. Edited by W. Beiglböck, A. Böhm and E. Takasugi. XIII, 540 pages. 1979.

Vol. 95: Quasi One-Dimensional Conductors I. Proceedings, 1978. Edited by S. Barišić, A. Bjeliš, J. R. Cooper and B. Leontič. X, 371 pages. 1979.

Vol. 96: Quasi One-Dimensional Conductors II. Proceedings 1978. Edited by S. Barišić, A. Bjeliš, J. R. Cooper and B. Leontić. XII, 461 pages. 1979.

Vol. 97: Hughston, Twistors and Particles. VIII, 153 pages. 1979.

Vol. 98: Nonlinear Problems in Theoretical Physics. Proceedings, 1978. Edited by A. F. Rañada. X, 216 pages. 1979.

Vol. 99: M. Drieschner, Voraussage – Wahrscheinlichkeit – Objekt. XI, 308 Seiten. 1979.

Vol. 100: Einstein Symposion Berlin. Proceedings 1979. Edited by H. Nelkowski et al. VIII, 550 pages. 1979.

Vol. 101: A. Martin-Löf, Statistical Mechanics and the Foundations of Thermodynamics. V, 120 pages. 1979.

Vol. 102: H. Hora, Nonlinear Plasma Dynamics at Laser Irradiation. VIII, 242 pages. 1979.

Vol. 103: P. A. Martin, Modèles en Mécanique Statistique des Processus Irréversibles. IV, 134 pages. 1979.

Vol. 104: Dynamical Critical Phenomena and Related Topics. Proceedings, 1979. Edited by Ch. P. Enz. XII, 390 pages. 1979.

Vol. 105: Dynamics and Instability of Fluid Interfaces. Proceedings, 1978. Edited by T. S. Sørensen. V, 315 pages. 1979.

Vol. 106: Feynman Path Integrals, Proceedings, 1978. Edited by S. Albeverio et al. XI, 451 pages. 1979.

Vol. 107: J. Kijowski, W. M. Tulczyjew, A Symplectic Framework for Field Theories. IV, 257 pages. 1979.

Vol. 108: Nuclear Physics with Electromagnetic Interactions. Proceedings, 1979. Edited by H. Arenhövel and D. Drechsel. IX, 509 pages. 1979.

Vol. 109: Physics of the Expanding Universe. Proceedings, 1978. Edited by M. Demiański. V, 210 pages. 1979.

Vol. 110: D. A. Park, Classical Dynamics and Its Quantum Analogues. VIII, 339 pages. 1979.

Vol. 111: H.-J. Schmidt, Axiomatic Characterization of Physical Geometry. V, 163 pages. 1979.

Vol. 112: Imaging Processes and Coherence in Physics. Proceedings, 1979. Edited by M. Schlenker et al. XIX, 577 pages. 1980.

Vol. 113: Recent Advances in the Quantum Theory of Polymers. Proceedings 1979. Edited by J.-M. André et al. V, 306 pages. 1980.

Vol. 114: Stellar Turbulence. Proceedings, 1979. Edited by D. F. Gray and J. L. Linsky. IX, 308 pages. 1980.

Vol. 115: Modern Trends in the Theory of Condensed Matter. Proceedings, 1979. Edited by A. Pekalski and J. A. Przystawa. IX, 597 pages. 1980.

Vol. 116: Mathematical Problems in Theoretical Physics. Proceedings, 1979. Edited by K. Osterwalder. VIII, 412 pages. 1980.

Vol. 117: Deep-Inelastic and Fusion Reactions with Heavy Ions. Proceedings, 1979. Edited by W. von Oertzen. XIII, 394 pages. 1980.

Vol. 118: Quantum Chromodynamics. Proceedings, 1979. Edited by J. L. Alonso and R. Tarrach. IX, 424 pages. 1980.

Vol. 119: Nuclear Spectroscopy. Proceedings, 1979. Edited by G. F. Bertsch and D. Kurath. VII, 250 pages. 1980.

Vol. 120: Nonlinear Evolution Equations and Dynamical Systems. Proceedings, 1979. Edited by M. Boiti, F. Pempinelli and G. Soliani. VI, 368 pages. 1980.

Vol. 121: F. W. Wiegel, Fluid Flow Through Porous Macromolecular Systems. V, 102 pages. 1980.

Vol. 122: New Developments in Semiconductor Physics. Proceedings, 1979. Edited by F. Beleznay et al. V, 276 pages. 1980.

Vol. 123: D. H. Mayer, The Ruelle-Araki Transfer Operator in Classical Statistical Mechanics. VIII, 154 pages. 1980.

Vol. 124: Gravitational Radiation, Collapsed Objects and Exact Solutions. Proceedings, 1979. Edited by C. Edwards. VI, 487 pages. 1980.

Vol. 125: Nonradial and Nonlinear Stellar Pulsation. Proceedings, 1980. Edited by H. A. Hill and W. A. Dziembowski. VIII, 497 pages. 1980.

Vol. 126: Complex Analysis, Microlocal Calculus and Relativistic Quantum Theory. Proceedings, 1979. Edited by D. Iagolnitzer. VIII, 502 pages. 1980.

Vol. 127: E. Sanchez-Palencia, Non-Homogeneous Media and Vibration Theory. IX, 398 pages. 1980.

Vol. 128: Neutron Spin Echo. Proceedings, 1979. Edited by F. Mezei. VI, 253 pages. 1980.

Vol. 129: Geometrical and Topological Methods in Gauge Theories. Proceedings, 1979. Edited by J. Harnad and S. Shnider. VIII, 155 pages. 1980.

Vol. 130: Mathematical Methods and Applications of Scattering Theory. Proceedings, 1979. Edited by J. A. DeSanto, A. W. Sáenz and W. W. Zachary. XIII, 331 pages. 1980.

Vol. 131: H. C. Fogedby, Theoretical Aspects of Mainly Low Dimensional Magnetic Systems. XI, 163 pages. 1980.

Vol. 132: Systems Far from Equilibrium. Proceedings, 1980. Edited by L. Garrido. XV, 403 pages. 1980.

Vol. 133: Narrow Gap Semiconductors Physics and Applications. Proceedings, 1979. Edited by W. Zawadzki. X, 572 pages. 1980.

Vol. 134: γγ Collisions. Proceedings, 1980. Edited by G. Cochard and P. Kessler. XIII, 400 pages. 1980.

Vol. 135: Group Theoretical Methods in Physics. Proceedings, 1980. Edited by K. B. Wolf. XXVI, 629 pages. 1980.

Lecture Notes in Physics

159

Erhard Seiler

Gauge Theories as a Problem of Constructive Quantum Field Theory and Statistical Mechanics

Springer-Verlag
Berlin Heidelberg GmbH 1982

Author
Erhard Seiler
Max-Planck-Institut für Physik
Postfach 401212, D-8000 München 40

ISBN 978-3-662-13520-4

Library of Congress Cataloging in Publication Data
Seiler, Erhard G., 1942- Gauge theories as a problem of constructive quantum field theory and statistical mechanics. (Lecture notes in physics; 159) Bibliography: p. Includes index.
1. Gauge fields (Physics) 2. Quantum field theory. 3. Statistical mechanics. I. Title. II. Series.
QC793.3.F5S44 1982 530.1'43 82-7343
ISBN 978-3-662-13520-4 ISBN 978-3-540-39245-3 (eBook)
DOI 10.1007/978-3-540-39245-3

Originally published by Springer-Verlag Berlin Heidelberg New York in 1982

2153/3140-543210

TABLE OF CONTENTS

Preface

These notes are based on a series of lectures I gave in Lausanne and Geneva in January and February of 1981 in the framework of the "Troisième Cycle de la Physique en Suisse Romande". A preliminary version of these notes was preprinted by the "Troisième Cycle". Compared to that version numerous changes and corrections have been made. In particular an erroneous proof of the combinatorial Corollary 3.9 has been replaced by a simpler and (I hope) correct one. I am grateful to many people, in particular Christian Borgs, Krzysztof Gawędzki, Konrad Osterwalder and Simon Ruijsenaars for pointing out various errors in the earlier version and for suggesting some improvements. I am also much indebted to Ricardo Neves da Silva for his help in preparing the manuscript.

Most of all I am indebted to the people who collaborated with me in the rigorous study of gauge quantum field theories, namely David Brydges, Jürg Fröhlich and Konrad Osterwalder. I also owe a great deal of my understanding of high- and low-temperature expansions to discussions with them; the third section would look rather horrible, I think, without their help (maybe it still does, but that is not their fault).

Furthermore my thanks go to Jean-Pierre Eckmann, Gérard Wanders and the "Commission Scientifique du Troisième Cycle de la Physique en Suisse Romande" for their kind invitation that made these lectures possible.

INTRODUCTION

The construction of quantized gauge theories is still in an embryonic stage in spite of their universally acknowledged importance in contemporary particle physics. In the meantime one is trying to obtain insight into the physical content of the theory (assuming its existence) by other methods. First there is renormalized perturbation theory which is quite well developed [1,2,3]. For the analysis of non-perturbative phenomena such as the presumed confinement of quarks the study of semiclassical approximations to the theory has enjoyed rather great popularity. Its successes, in my opinion , so far lie more in the realm of the classical theory, which is a beautiful mathematical subject in its own right, than in the understanding of the quantized theory; this is not meant to deny that this approach has produced many fascinating ideas and some insights (see [4] and references given there). In any case, in these lectures I will not be concerned with these two approaches. Instead I will concentrate in the first half on a different nonperturbative method : The lattice approximation. Since the invention of lattice gauge theories by Wegner [5] and in particular since Wilson [6] (apparently unaware of Wegner's paper) introduced them into particle physics as a way to analyze specific physical questions, there has been tremendous activity in that field and it will be impossible to give a complete review of all the results, quasiresults and ideas that have been produced. I will therefore concentrate on rigorously established results of physical relevance; all of them are the fruit of the observation that the lattice approximation makes gauge theories into some kind of classical statistical mechanics and interesting questions can be tackled with the highly developed machinery of that field of research.

Lately some people advocated taking this statistical mechanics aspect more literally and using lattice theories as models for the statistical mechanics of defects in ordered media [7]. This might yet turn out to be the most realistic area of application of these models as the particle physics interpretation of the results achieved so far still has a somewhat metaphoric character.

But the lattice approach is also important for another reason : It provides a tool for the construction of continuum models. In spite of a certain awkwardness it has provided the most successful road for the construction of continuum theories. In the second part of these lectures I will describe in some detail the arduous journey along this road to the construction of the two-dimensional Higgs model (Landau - Ginzburg model) and two-dimensional quantum electrodynamics (QED_2). I will also at least briefly discuss some other programs that are under way or, in some cases, have been completed some time ago. Even though for the Higgs model Wightman's axioms have been verified, it seems that in particular for confining theories they do not provide the most natural framework. So at the end of these lectures I will discuss a framework that is dealing with fields not "living on points" but rather on curves or loops; thereby it is possible to avoid state spaces with "indefinite metric" that are used in the perturbative approach and also in the axiomatic framework proposed for instance by Strocchi [8].

The geometric aspects of gauge theories will not play a central rôle in these lectures. But since it is useful to have the geometric interpretation in the back of one's mind and since I will sometimes use the geometric language, I give a very brief and informal description of the concepts, mainly intended as a glossary in an appendix. For a more detailed discussion see [88,89] or the beautiful review by Eguchi, Gilkey and Hanson [4].

I. LATTICE GAUGE THEORIES

Lattice gauge theories were first studied by Wegner [5] under a different name (for the gauge group $\mathbb{Z}_2$) . He was interested in generalizations of the Ising model which possess phase transitions without a local order parameter. In this paper the relevant "order" and "disorder" observables which are today known as "Wilson loop" and "'t Hooft loop" are introduced and their behavior in different regimes (area- or perimeter law, respectively) is studied. A few years later Wilson [6] introduced a rather general class of lattice gauge theories in order to understand the permanent confinement of quarks. He formulated what is today known as "Wilson's criterion" : A gauge theory confines quarks if the appropriate loop observable ("Wilson loop") obeys the area law. We will discuss the meaning and proof of this criterion as well as its limitations and possible alternatives below.

There are two review articles on the subject, one by Drouffe and Itzykson [9], the other one by Kogut [10]. Both contain a lot of useful information and are extremely valuable introductions into the subject but they do not cover most of the more mathematically rigorous work in the field. In some sense my presentation here will be complementary to theirs.

I will restrict the discussion to the properties of lattice gauge theories at temperature zero which simply means that the full thermodynamic limit will be taken instead of keeping a finite size in time direction with periodic boundary conditions. There is interesting physics in these finite temperature gauge theories, in particular the presumed existence of a critical temperature in 4 dimensions above which quarks cease to be confined [84, 85] , but in the interest of keeping the size of these notes manageable I will not go into this subject.

The major physical problem which can be studied in the framework of lattice gauge theories is still the one that originally motivated Wilson [6] to invent them: The permanent confinement of quarks at zero temperature. In spite of all the effort put into the analysis of this problem it can still not yet be said that confinement has been proven, not even in the sense of Wilson's criterion, for four-dimensional non-abelian models of arbitrary coupling. But many partial results have been obtained and a physical picture has emerged that gives support to the idea that the vacuum of a confining theory resembles a magnetic superconductor and thereby squeezes the color electric field by a "dual" Meissner effect into tubes between the charges, thus producing a force between them that is essentially independent of their distance. It turns out that this confinement mechanism only applies to charges that transform nontrivially under the center of the gauge group. Therefore for other nontrivial charges (for instance those having the quantum numbers of gluons) a different mechanism is believed to operate that screens their charge; in the end all physical states are believed to be color neutral. This mechanism also provides an explanation for the "saturation of forces" that manifests itself for instance in the fact that there are no confining forces between objects made from three quarks.

The Wilson criterion for confinement only talks about the pure Yang-Mills theory; there is no such simple criterion for models that contain matter fields transforming nontrivially under the center of the gauge group, such as Quantum Chromodynamics (QCD). It should be stressed that confinement means more than just the absence of states with nonvanishing color charge : It would be a disaster for the confinement dogma if quarks managed to screen their color and thereby escaped the confining force.

For a good discussion of the physical picture of confinement and screening I recommend [86]. We will return to the issues raised here at various points in these lectures.

1. THE SCHEME OF LATTICE GAUGE THEORIES

The heuristic background which should be kept in mind is the euclidean version of the Feynman path integral that has led to considerable successes in Constructive Quantum Field Theory (see [11, 12, 80] for a review and references).

There the idea is to use the classical action $S(\phi)$, where ϕ symbolizes the fields of the model, to construct a probability measure on the space of "field configurations" by the prescription

$$d\mu(\phi) = \frac{1}{Z} e^{-S(\phi)} \prod_x d\phi(x)$$

Here $\prod_x d\phi(x)$ stands for the (nonexisting) Lebesgue measure on the fields and Z is a normalization factor to be chosen to ensure $\int d\mu = 1$. Lattice versions of this formula have been employed in the case of scalar field theories with great success since they allowed to use methods of statistical mechanics [13, 14]; the continuum limit could actually be controlled in 2 and 3 dimensions [13, 14].

In the case of a gauge field A_μ , its geometrical meaning should be remembered before introducing a lattice. A_μ , as explained in the appendix,are the relevant components of a connection form in a principal bundle, expressed in a particular coordinate system (which includes a local trivialization of the bundle). $A \equiv A_\mu dx^\mu$ is a 1-form with values in the Lie algebra $\mathfrak{g}$ of the chosen gauge group G (assumed to be a compact Lie group) telling you in which way you have to move in internal symmetry space when you move in a certain direction in the base space (= euclidean space-time). It induces (for a given local trivialization ≡ "gauge") a map from curves C_{xy} starting at x and ending at y into the gauge group G itself by the well known prescription

$$g(C_{xy}) \equiv P \exp \int_{C_{xy}} A_\mu dx^\mu \tag{1.1}$$

where P stands for "path ordering". The right hand side is the well-known product integral (cf. [15]) and the physicist's notation employed here conveys its meaning very clearly.

Going from euclidean space-time $\mathbb{R}^d$ to a simple cubic lattice $\varepsilon\,\mathbb{Z}^d$ (or any other lattice if that is desired) we understand by the gauge field now a map from the oriented "links" or nearest neighbor pairs $\langle xy \rangle$ of the lattice into the group G

$$\langle xy \rangle \mapsto g_{xy} \in G \tag{1.2}$$

obeying

$$g_{xy} = g_{yx}^{-1} \tag{1.3}$$

Sometimes it will be useful to think of g_{xy} as arising from an underlying continuum gauge field by a prescription such as (1.1).

By our construction it is clear what we understand by a gauge transformation of $\langle xy \rangle \mapsto g_{xy}$: It will be described by a map from the sites into G

$$x \mapsto h_x \tag{1.4a}$$

and change $\quad g_{xy}$ into $h_x g_{xy} h_y^{-1}$. (1.4b)

The next goal is to define a lattice version of the continuum action of the gauge field which is given by

$$S_{Y.M.}(A) \equiv -\frac{1}{2g_o^2} \int \mathrm{Tr}(F \wedge {*F})$$
$$\equiv -\frac{1}{4g_o^2} \int \mathrm{Tr}\, F_{\mu\nu} F_{\mu\nu} \equiv \frac{1}{2g_o^2} \|F\|_2^2 \tag{1.5}$$

where $F \equiv dA + \frac{1}{2}[A,A]$ is the curvature 2-form associated with A (taking values in $\mathfrak{g}$) and the trace is to be taken in any locally faithful representation; $*F$ is the Hodge dual of F .

Wilson's lattice version of (1.5) is obtained as follows : Let χ be any character of G belonging to a locally faithful representation. The lattice gauge field associates to any closed oriented loop C a conjugacy class $[g_C]$ of G by simply multiplying the group elements g_{xy} corresponding to the links in C , starting at any point and using the order given by the path C . In particular the elementary squares ("plaquettes") P (or rather their boundaries) of the lattice will be mapped into conjugacy classes $[g_{\partial P}]$.

We define now :

$$S_{Y.M.W.}(\{g_{xy}\}) \equiv -\frac{1}{2g_o^2}\sum_P \chi(g_{\partial P}) \ . \qquad (1.6)$$

where the sum is over all oriented plaquettes. The justification for this definition lies in the fact that

$$\chi(g_{\partial P}) = \chi(\mathbf{1}) + \frac{\varepsilon^4}{2}\chi(F_{\mu\nu}F_{\mu\nu}) + O(\varepsilon^6) \qquad (1.7)$$

whenever the lattice gauge field arises from a (sufficiently smooth) continuum gauge field.

Of course (1.6) is by no means unique; there are many alternatives that formally correspond to the same continuum limit. It is an important question whether the true, nonformal continuum limit will be independent of the choice of lattice approximation; the answer seems actually to be "no" in general as we will see below. We will take a pragmatic point of view with regard to this question and choose our lattice action by convenience; the choice should be justified by its success.

Note that even (1.6) is not unambiguously defined because it depends on the choice of the character χ . This will turn out to be very relevant in connection with the confinement problem. In the standard case of $G = SU(N)$ we will insist on using the fundamental representation because it represents the center $\mathbb{Z}_N$ faithfully. It is curious that the formal continuum theory does not have this dependence. But it seems that the continuum limit of the lattice theory will depend crucially on the way in which the center is represented.

We want to mention one alternative to (1.6) which presumably has the same continuum limit, which we call "Villain-Polyakov action" (because it generalizes Villain's form of the plane rotator model [16]) and which was studied by Drouffe [17].

For a compact semisimple Lie group G it is given by

$$\exp(-S_{Y.M.V.})(\{g_{xy}\}) \equiv \prod_P \sum_i \exp\left(-\frac{g_o^2}{2}C_\tau\right)\chi_\tau(g_{\partial P})d_\tau \qquad (1.8)$$

Here the sum is over all inequivalent irreducible (unitary) representations U_τ of G , $\chi_\tau(g) = \mathrm{tr}\, U_\tau(g)$, $d_\tau = \chi_\tau(\mathbf{1})$ and C_τ is the eigenvalue of the quadratic Casimir operator in the representation τ , i.e. if $X_1,\dots X_n$ is an orthonormal basis of the Lie algebra $\mathcal{g}$ with respect to the Killing form $k(X,Y) \equiv \mathrm{tr}\, U_{adj}(X)U_{adj}(Y)$ then

$$C_\tau = -\sum_{i=1}^n \mathrm{tr}\, U_\tau(X_i)^2$$

This means that $\sum_\tau \exp(-\frac{g_o^2}{2}C_\tau)\chi_\tau(gh^{-1})$ is the kernel of the heat operator $\exp\frac{1}{2}g_o^2\Delta_B$ where Δ_B is the Laplace-Beltrami operator with respect to the Killing metric on G . For non-semisimple G we may define $S_{Y.M.V.}$ by using any invariant metric on G and using the corresponding heat kernels.

Note the formal continuum limit :

$$\log \frac{\sum_\tau \exp(-\frac{g_o^2}{2} C_\tau) d_\tau \chi_\tau(g_{\partial P})}{\sum_\tau \exp(-\frac{g_o^2}{2} C_\tau) d_\tau \chi_\tau(\mathbb{1})}$$

$$= \frac{\varepsilon^4}{2} \frac{\sum_\tau \exp(-\frac{g_o^2}{2} C_\tau) d_\tau \chi_\tau(F_{\mu\nu} F_{\mu\nu})}{\sum_\tau \exp(-\frac{g_o^2}{2} C_\tau) d_\tau \chi_\tau(\mathbb{1})} + O(\varepsilon^6)$$

$$= \frac{\varepsilon^4}{2} \frac{\sum_\tau \exp(-\frac{g_o^2}{2} C_\tau) d_\tau^2 C_\tau}{\sum_\tau \exp(-g_o^2 C_\tau) d_\tau^2} \chi_{adj}(F_{\mu\nu} F_{\mu\nu}) + O(\varepsilon^6)$$

$$\sim \frac{\varepsilon^4}{2} \begin{cases} \frac{1}{g_o^2 d_{adj}} \chi_{adj}(F_{\mu\nu} F_{\mu\nu}) & \text{for } g_o^2 \text{ small} \\ C_1 d_1 \exp(-\frac{g_o^2}{2} C_1) \frac{1}{d_{adj}} \chi_{adj}(F_{\mu\nu} F_{\mu\nu}) & \text{for } g_o^2 \text{ large} \end{cases}$$

where χ_{adj} stands for the trace in the adjoint representation, d_{adj} is the dimension of the adjoint representation, C_1 is the smallest Casimir eigenvalue and d_1 the dimension of the corresponding representation space.

It is fairly straighforward to introduce matter fields into the system. They come in two basic varieties :

(1) Scalar Higgs fields

(2) Spinor fields (corresponding to quarks and/or leptons).

The Higgs fields as well as the spinor fields "live" on the sites of the lattice; they are coupled to the gauge fields by a lattice version of the standard "minimal coupling"; the Higgs fields will in addition have some suitable self-coupling. To avoid too cluttered formulae we set now $\varepsilon = 1$; it is easy to reinsert ε when needed.

A <u>lattice Higgs field configuration</u> is a map ϕ from the sites x of the lattice into some finite dimensional unitary (or euclidean) vector space V_H carrying a unitary (or orthogonal) representation U_H of the gauge group G .

$$\phi : x \mapsto \phi(x) \in V_H$$

Sometimes we will restrict the values of ϕ to lie in a sphere of a fixed radius R in U_H , i.e. $\|\phi(x)\| = R$ for all x .

The lattice Higgs action is

$$S_H(\{\phi(x)\},\{g_{xy}\}) \equiv -\lambda/2\Big(\sum_{\langle xy\rangle} (\phi(x),U_H(g_{xy})\phi(y))\Big) + \sum_x V(\|\phi(x)\|) . \tag{1.9}$$

Here V is an even polynomial of degree ≥ 4 with positive leading coefficient and the first sum is over oriented links $\langle xy\rangle$. To become truly "Higgsian", V will have to have a deep absolute minimum far away from zero.

A lattice spinor field configuration is a map ψ from the sites of the lattice into a subset of the set of orthonormal frames of a fermionic vector space V_F .

V_F has the following structure :

$$V_F = V_S \otimes V_G \tag{1.10}$$

where V_S ("spin space") is a representation space for the Dirac-Clifford algebra defined by

$$\gamma_i\gamma_j + \gamma_j\gamma_i = 2\delta_{ij} \qquad (i,j = 0,1,\dots d-1) \tag{1.11}$$

through hermitean matrices. V_G ("gauge space") is a unitary space carrying a unitary representation U_F of the gauge group G .

The field configuration ψ is required to "respect" the decomposition of V_F into tensorial factors, i.e. we require it to be of the form

$$\begin{aligned} \psi : x \mapsto \{\psi_{\alpha a}(x)\}_{\substack{\alpha=1,\dots,\dim V_S \\ a=1,\dots,\dim V_G}} \\ \psi_{\alpha a}(x) = e_\alpha(x) \otimes f_a(x) , \quad e_\alpha \in V_S , \; f_a \in V_G \end{aligned} \tag{1.12}$$

where α labels "spin" and a "internal" degrees of freedom.

Next we pick an antiunitary map I from V_G into an isomorphic copy $\overline{V}_G$; correspondingly we will have a field configuration $\overline{\psi} = (\mathbb{1} \times I)\psi$:

$$\overline{\psi} : x \mapsto \{\overline{\psi}_{\alpha a}(x)\}_{\alpha,a} \tag{1.13}$$

On $\overline{V}_G$ a unitary representation $U_{\overline{F}} = IU_F I^{-1}$ of G which is conjugate to U_F is naturally induced. Note that if

$$U_F \psi_a = (U_F)_{ab}\psi_b \quad , \text{ then}$$

$$U_{\overline{F}}\overline{\psi}_a = IU_F\psi_a = (\overline{U}_F)_{ab}\overline{\psi}_b = (U_F^{-1})_{ba}\overline{\psi}_b \quad .$$

We now regard the orthonormal vectors $\{\psi_{\alpha a}(x), \overline{\psi}_{\alpha a}(x)\}$ as generators of the exterior algebra over $\bigotimes_x (V_F \oplus \overline{V}_F)$ where $\overline{V}_F = V_S \otimes \overline{V}_G$; thereby we can construct functions from spinor field configurations into that exterior algebra by forming polynomials (in the sense of exterior multiplication). For instance the fermionic action is

$$\begin{aligned} &S_F(\{\psi_{\alpha,a}(x), \overline{\psi}_{\alpha,a}(x)\}, \{g_{xy}\}) \equiv \\ &\equiv \frac{\kappa}{2} \sum_{<xy>} \overline{\psi}_{\alpha a}(x)\Gamma^{xy}_{\alpha\beta}(U_F)_{ab}\psi_{\beta b}(y) + \\ &+ \frac{1}{2}\sum_x \overline{\psi}_{\alpha a}(x)\Gamma_{\alpha\beta}\psi_{\beta a}(x) \end{aligned} \tag{1.14}$$

The naïve choice $\Gamma \equiv M$ and

$$\Gamma^{xy} \equiv \begin{cases} \gamma_\mu & \text{if } <xy> \text{ points in the positive } \mu\text{-direction} \\ -\gamma_\mu & \text{if } <yx> \text{ points in the positive } \mu\text{-direction} \end{cases}$$

leads to a proliferation of fermion degrees of freedom in the continuum limit [18,51,92].

This is avoided by the following two-parameter family of choices

$$\Gamma_\theta \equiv M - rde^{i\theta\gamma_5}$$

$$\Gamma^{xy}_\theta \equiv \begin{cases} re^{i\theta\gamma_5} + \gamma_\mu & \text{if } <xy> \text{ points in the positive } \mu\text{-direction} \\ re^{i\theta\gamma_5} - \gamma_\mu & \text{if } <yx> \text{ points in the positive } \mu\text{-direction} \end{cases}$$

$(0 < r \leq 1)$.

For $\theta = 0$, $r = 1$ this is Wilson's action [18] , for $\theta = \frac{\pi}{2}$, $r = 1$, it is the action proposed in [19]. (γ_5 is the conventional notation for a hermitian matrix obeying

$$\gamma_5^2 = 1 \ , \quad \gamma_5\gamma_i + \gamma_i\gamma_5 = 0 \quad \text{for} \quad i = 0,1,\dots,d-1$$

and it exists if the dimension of the chosen representation of the Clifford algebra is high enough).

Remark : We want to stress the (well known) fact that with this choice (1.14) is no longer invariant under the chiral transformation

$$\psi \to e^{i\alpha\gamma_5}\psi \quad , \quad \overline{\psi} \to \overline{\psi} e^{i\alpha\gamma_5}$$

even for $M = 0$; this leads to the well known axial anomalies in the continuum limit (see [81]).

A related fact is the following : The continuum limit formally seems to be θ independent. A more careful analysis [95] (see Section 5) shows, however, that often the angle θ is remembered in the continuum. The corresponding θ-dependent states are known as "θ-vacua"; in a different context we will encounter them in Section 4a .

Let us now assume that we are working on a finite chunk Λ of our lattice. This makes our somewhat formal definitions $S_{Y.M.}, S_H, S_F$ sensible. We define a field algebra A_Λ as the Grassmann algebra generated by $\{\psi_{\alpha a}(x) \ , \ \overline{\psi}_{\alpha a}(x)\}_{x\in\Lambda}$ with coefficients that are bounded continuous functions of the bosonic fields $\{g_{xy}\}$, $\{\phi(x)\}$. A_Λ may be viewed as a map from field configurations into the Grassmann algebra

$$G_\Lambda \equiv \Lambda((\bigotimes_{x\in\Lambda} V_F) + (\bigotimes_{x\in\Lambda} V_F))$$

On A_Λ we define a norm $||\cdot||$ as follows : The exterior algebra G_Λ decomposes into a direct sum of homogeneous linear subspaces :

$$G_\Lambda = \sum_{n\geq 0} \Lambda^n((\bigoplus_x V_F) + (\bigoplus_x \overline{V}_F)) \equiv \sum_{n\geq 0} G_\Lambda^{(n)}$$

Each of the subspaces $G_\Lambda^{(n)}$ inherits a positive definite inner product and a norm $||\cdot||$ from V_F ; for $A \in G$ we have the decomposition $A = \sum A^{(n)}$ with $A^{(n)} \in G_\Lambda^{(n)}$ and we define $||A|| \equiv \sum_{n\geq 0} ||A^{(n)}||$. With this definition we have $||AB|| \leq ||A||\ ||B||$.

To obtain a norm on A_Λ we only have to replace the complex numbers which play the rôle of scalars in G_Λ by the bounded continuous functions of the bosonic fields, normed by the sup-norm. So if $F = \sum_{n\geq 0} F^{(n)}$ is the decomposition of $F \in A_\Lambda$ into homogeneous elements in the fermion variables, we define

$$\|F\| \equiv \sum_{n\geq 0} \sup_{\{\phi(x),g_{xy}\}} \|F^{(n)}\|$$. Again we have $\|FG\| \leq \|F\| \, \|G\|$.

Finally an expectation value is defined as a linear functional $\langle\cdot\rangle$ on A_Λ as follows (based on [58]): First we define a "decoupled" expectation $\langle\cdot\rangle_o$ as a linear functional on A_Λ by requiring

(a) $\langle P\rangle_{\Lambda,o} = 0$ whenever P is an element of less than maximal degree in the exterior algebra

(b) $$\langle F(\{\phi(x)\},\{g_{xy}\}) \bigwedge_{\substack{\alpha,a \\ x\in\Lambda}} (\bar{\psi}_{\alpha,a}(x)\psi_{\alpha,a}(x))\rangle_{\Lambda,o} = \int d\mu^o_\Lambda F$$

where $d\mu^o_\Lambda$ is the following positive measure

$$d\mu^o_\Lambda = \prod_{\langle xy\rangle\in\Lambda\times\Lambda} dg_{xy} \prod_{x\in\Lambda} e^{-V(|\phi(x)|)} d\phi(x)$$

with dg (normalized) Haar measure on G and $d\phi$ Lebesgue measure in V_H . It is easy to see that $|\langle F\rangle_o| \leq \|F\|$ const. Finally we define for $F \in A_\Lambda$:

$$\langle F\rangle_\Lambda \equiv \frac{1}{Z_\Lambda} \langle F\, e^{-\mathring{S}_H - S_{Y.M.} - S_F}\rangle_{\Lambda,o} \ , \quad Z_\Lambda \equiv \langle e^{-\mathring{S}_H - S_{Y.M.} - S_F}\rangle_{\Lambda,o} \ .$$

$\mathring{S}_H$ stands for the Higgs action with V replaced by 0 .

Note that both the functional $\langle\cdot\rangle_{\Lambda,o}$ and the functional $\langle\cdot\rangle_\Lambda$ are "gauge invariant" in the sense that they do not change if we replace

g_{xy} by $h_x g_{xy} h_y^{-1}$

$\phi(x)$ by $U_H(h_x)\phi(x)$

$\psi(x)$ by $U_F(h_x)\psi(x)$

$\bar{\psi}(x)$ by $\bar{\psi}(x)U_F(h_x^{-1})$

simultaneously.

The appropriate modifications needed if some of the fields are absent are obvious. A final remark : For the lattice theories we may relax the requirement that G is a compact Lie group; G may be any compact group.

So what we have gotten by our lattice approximation looks a lot like statistical mechanics of more conventional lattice systems, except for the somewhat unfamiliar Grassmann nature of the fermion fields. We will see that a typical problem of statistical mechanics, namely the thermodynamic limit, will be tackled by methods lifted right out of the conventional toolkit; the same is true of a number of other questions of interest.

It should also be pointed out that we "integrate" over all field configurations, including gauge equivalent ones. Unlike the formal continuum case there is no need "to fix the gauge" introducing strange beings like Faddeev-Popov ghosts - even though it could be done if desired (cf. [19, 20]).

2. FUNDAMENTAL PROPERTIES

Here we will describe and prove a few basic properties of lattice gauge theories that do not involve more refined methods such as expansions or "integral equations".

a) Osterwalder-Schrader Positivity and Consequences.

This property, often known as reflection positivity, has been known and used for certain lattice systems for quite a while [21,22]. It is not to be confused with the positivity of the bosonic measure $d\mu_\Lambda$ introduced above (which comes under the heading "Symanzik-Nelson positivity" in the context of Euclidean quantum field theory). O.S. positivity is true even under the inclusion of fermions; it is a useful property for "technical" purposes, but its fundamental meaning comes from the fact that it allows to construct a quantum mechanical state space with positive definite scalar product; this property will also be inherited by any reasonable continuum limit. O.S. positivity for lattice gauge theories was proven in [19], cf. also [23].

We will restrict ourselves to the following two situations :

(1) Our finite lattice $\Lambda \subset \mathbb{Z}^d$ is symmetric with respect to a hyperplane ("t = 0 hyperplane") lying halfway between lattice hyperplanes:

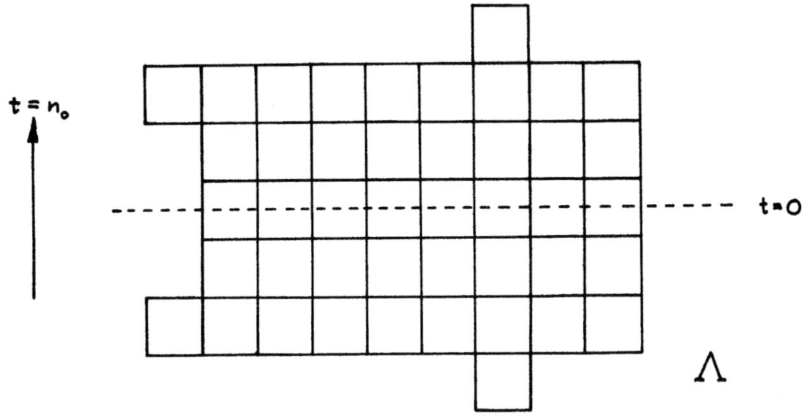

(2) Our finite lattice Λ is wrapped around a torus (or cylinder); more specifically we will identify points

$$(n_o, n_1, \dots, n_{d-1})$$

and

$$(n_o + 2N, n_1, \dots, n_{d-1}) \quad \text{etc.}$$

If fermions are present, they will have to obey so-called antiperiodic boundary conditions

$$\psi(n_o, n_1, \dots, n_{d-1}) = -\psi(n_o + 2N, n_1, \dots, n_{d-1}) \tag{2.1}$$

while the bosonic fields will obey ordinary periodic b.c. (If (2.1) sounds incompatible with the identification proposed just before, you may replace it by ordinary periodic b.c. and a modification in the fermionic action : The sign of the terms coupling positive and negative times "on the back", i.e. involving "times" N and $-N$ has to be flipped. It is a little exercise to show that equivalently one may choose any other time layer).

In both situations (1) and (2) there is a natural decomposition

$$\Lambda = \Lambda_+ \cup \Lambda_- \ ; \ \Lambda_+ \cap \Lambda_- = \emptyset \tag{2.2}$$

and a map r ("reflection at the $t = 0$ plane")

$$r : \Lambda_\pm \to \Lambda_\mp \tag{2.3}$$

This reflection induces an antilinear map θ of the field algebras $\mathcal{A}_\Lambda$ belonging to Λ by requiring

(a) $\theta F(\{\phi(x)\}) = \overline{F(\{\phi(rx)\})}$

$\theta G(\{g_{xy}\}) = \overline{G(\{g_{rx,ry}\})}$

(b) $\theta\psi(x) = \overline{\psi}(rx)\gamma_o$

$\theta\overline{\psi}(x) = \gamma_o\psi(rx)$

(c) $\theta(AB) = (\theta B)(\theta A)$

It is clear how requirements (a), (b), (c) uniquely specify θ as an antilinear map from A_Λ to A_Λ with the additional property

$$\theta \restriction A_{\Lambda_\pm} = A_\Lambda \tag{2.4}$$

We can now formulate

Theorem 2.1 : (O.S. positivity, first proven in [19]; but the proof given there contains a slight inaccuracy). For $F \in A_{\Lambda_+}$,

$$\langle F\theta F\rangle_\Lambda \geq 0$$

Proof : First assume that F is either even or odd. Note that

$$\langle F\theta F\rangle_{o,\Lambda} = \langle F\rangle_{o,\Lambda_+}\langle\theta F\rangle_{o,\Lambda_-} = |\langle F\rangle_{o,\Lambda_+}|^2 \geq 0 \ .$$

The elements of the form $\sum_{i=1}^{n} c_i F_i \theta F_i$, $c_i \geq 0$, $F_i \in A_{\Lambda_+}$, F_i even or odd (i=1,...,n) form a "multiplicative cone" P , i.e. products (and of course linear combinations with positive coefficients) are again of that form.

Therefore $\exp(F\theta F)$ will belong to that cone for $F \in A_{\Lambda_+}$, F even or odd. Now

$$-S \equiv -(\overset{\circ}{S}_H + S_{Y.M.} + S_F)$$

can be written as the sum of three terms :

$$-S = -S_+ - \theta S_+ - S_c$$

with $-S_+ \in A_{\Lambda_+}$. $-S_+$ collects all terms "living" entirely in Λ_+ ; $-\theta S_+$ all terms in Λ_- and the rest contains everything that couples Λ_+ and Λ_- . Now if $\exp(-S_c)$ belonged to the cone P we would be done because then $F\theta F\, e^{-S}$ would belong to P . Unfortunately $\exp(-S_c)$ is not in P as it stands, but we can use the gauge freedom to get it there : We postpone the integration over all g_{xy} with $x \in \Lambda_+$, $y \in \Lambda_-$ or vice versa. This corresponds to a "conditional expectation" $\langle\cdot\rangle_{\{g_{x,rx}\},\Lambda}$. We claim that for F that are either gauge invariant or do not depend on those "dangerous" variables $g_{x,rx}$,

$$\langle F\rangle_{\{g_{x,rx}\},\Lambda} = \langle F\rangle_\Lambda \tag{2.5}$$

(which justifies the term "conditional expectation"). The proof is by inspection : $\langle F\rangle_{\{g_{x,rx}\},\Lambda}$ has to be a gauge invariant function of its variables, but there is a gauge in which they are all equal to $\mathbb{1}$. So the left hand side of (2.5) is actually independent of $\{g_{x,rx}\}$ and the final integration over these variables has no effect. In particular we have

$$\langle F\rangle_\Lambda = \langle F\rangle_{\{g_{x,rx}=\mathbf{1}\},\Lambda} \ . \tag{2.6}$$

But the right hand side of this is simply obtained from the left hand side by replacing all the $g_{x,rx}$ in $-S_c$ by $\mathbf{1}$. $\exp(-S_c(\{g_{x,rx}=\mathbf{1}\}))$ belongs to P as can be seen by close inspection (see [92] for more details). Now we realize that the restriction requiring F to be even or odd is irrelevant since there are no cross terms. So the proof is complete. □

A direct consequence is the existence of a quantum mechanical Hilbert space $\mathcal{H}$: Let $\mathcal{N} \equiv \{F \in \mathcal{A}_{\Lambda_+} \mid \langle F\theta F\rangle_\Lambda = 0\}$. Then $\mathcal{A}_{\Lambda_+}/\mathcal{N}$ carries a positive definite scalar product given by $\langle F\theta F\rangle_\Lambda$. Its completion will be defined to be $\mathcal{H}$:

$$\mathcal{H} \equiv \overline{\mathcal{A}_{\Lambda_+}/\mathcal{N}} \ .$$

Another straightforward consequence is the existence of a positive transfer matrix : Let us assume that we can construct the thermodynamic limit at least in "time" direction and that it is time translation invariant. If then $F \in \mathcal{A}_{\Lambda_+}$, denote by $\tilde{T}F$ the same function but of the fields shifted by 2 units in positive time direction. Now clearly $\langle\tilde{T}^N F\theta F\rangle_\Lambda$ is bounded uniformly in N. By iterating the Schwarz inequality

$$|\langle\tilde{T}F\theta F\rangle_\Lambda| \leq \langle(\tilde{T}^2F)\theta F\rangle_\Lambda^{1/2}\langle F\theta F\rangle_\Lambda^{1/2} \leq$$

$$\leq \langle(\tilde{T}^{2^n}F)\theta F\rangle_\Lambda^{2^{-n}} \langle F\theta F\rangle_\Lambda^{1-2^{-n}} \ .$$

For $n \to \infty$ the first factor goes to 1, so we obtain

$$|\langle\tilde{T}F\theta F\rangle_\Lambda| \leq \langle F\theta F\rangle_\Lambda \tag{2.7}$$

This shows that $\widetilde{T}$ induces a well defined contraction T on the equivalence classes A_{Λ_+}/N and therefore on H . Also clearly $\langle (TF)\theta F\rangle_\Lambda \geq 0$, so we obtain $0 \leq T \leq 1$ on H .

Remarks :

1. Lüscher [24] has shown in a somewhat different (and more restricted) setting that actually $T > 0$.

2. In [23] a slightly more general version of O.S. positivity is proven which requires the introduction of some auxiliary fields ("half gauge fields").

A third well known consequence of O.S. positivity are the so-called chessboard bounds (see [22,25] and many references given in [23]). In our framework they can be stated as follows :

Let F_x be a "local" function of the fields. By this we mean that F_x may depend on the matter fields at site x . Furthermore let for a pair of sites x,y : $\sigma_j = (y_j - x_j)$. Also let r_j , θ_j be the reflections with respect to the hyperplane $x_j = 0$ (lying halfway between two lattice planes) and τ_{xy} the translation from $(\prod_{j=o}^{d-1} r_j^{\sigma_j})x$ to y . Let $F_{(xy)}$ be the following shifted and reflected copy of F_x :

$$F_{(xy)} \equiv \tau_{xy}((\prod_{j=o}^{\nu-1} \theta_j^{\sigma_j})F_x) \quad .$$

Then for (anti)periodic b.c. we have

Theorem 2.2 :

$$|\langle \prod_{x\in\Lambda} F_x \rangle_\Lambda| \leq \prod_{x\in\Lambda} \langle \prod_{y\in\Lambda} F_{(xy)} \rangle_\Lambda^{\frac{1}{|\Lambda|}}$$

The proof involves essentially an infinitely repeated application of Schwarz's inequality. This is systematized in [22,25]. □

Extensions to include functions of the gauge field are possible [23].

b) Some Observables and Their Meaning

O.S. positivity also allows to give physical interpretation to two types of popular observables : Wilson loops and 't Hooft loops (vortices, monopoles). We assume that the thermodynamic limit has been performed and so we have a non-negative transfer matrix.

First notice that the physical Hilbert space $\mathcal{H}$ is automatically gauge invariant : two elements of A_{Λ_+} that are related by a gauge transformation differ only by an element of the null space N because of the gauge invariance of the expectation (exercise!).

It is possible, however, to construct a larger Hilbert space $\tilde{\mathcal{H}}$ on which time independent gauge transformations act as nontrivial unitary operators. It is obtained as follows :

Let $\tilde{\mathcal{A}}_{\Lambda_+}$ be the algebra of functions of fields in Λ_+ in the "temporal gauge" (that means after setting all gauge fields in "time" direction equal to $\mathbf{1}$) . Expectations $\langle\cdot\rangle_{t.g.}$ in the temporal gauge are defined by also replacing the gauge fields in time direction in the action by $\mathbb{1}$. It is straightforward to see that

$$\langle\tilde{F}\theta\tilde{F}\rangle_{t.g.} \geq 0 \quad \text{for} \quad \tilde{F} \in \tilde{\mathcal{A}}_{\Lambda_+} \quad ,$$

so we can define $\tilde{\mathcal{H}} \equiv \overline{\tilde{\mathcal{A}}_{\Lambda_+}/N}$ as before.

Finally we note that if F is a gauge invariant element of A_Λ and $\tilde{F}$ the corresponding temporal gauge object, $\langle F\rangle = \langle\tilde{F}\rangle_{t.g.}$. This leads to a natural imbedding of $\mathcal{H}$ in $\tilde{\mathcal{H}}$.

Now consider a state $\psi \in \tilde{\mathcal{H}}$ and let $V(h_x)$ be the unitary operator induced on $\tilde{\mathcal{H}}$ by the time-independent gauge transformation symbolized by h_x (x is a fixed arbitrary point in the "spatial" lattice $\mathbb{Z}^{d-1}$) . $V(h_x)\psi$ has a Fourier decomposition into irreducible components :

$$V(h_x)\psi = \sum_\tau V_\tau(h_x)\psi_\tau$$

where

$$\psi_\tau = \int dh_x V(h_x)\chi_\tau(h_x^{-1})d_\tau\psi$$

$$V_\tau(h_x) = \int dg_x V(g_x)\chi_\tau(g_x^{-1}h_x)d_\tau$$

We say that ψ_τ has charge τ at the point x .

The strange thing about $\tilde{\mathcal{H}}$ is the fact that the decomposition obtained in this way

$$\tilde{\mathcal{H}} = \bigoplus_{n=o}^{\infty} \bigoplus_{\substack{x_1,\dots,x_n \\ \tau_{x_1},\dots,\tau_{x_n}}} \mathcal{H}_{\tau_{x_1},\dots,\tau_{x_n}} \qquad \text{(sum only over nontrivial } \tau\text{'s)}$$

is a decomposition into infinitely many superselection sectors characterized by the location of the charges and invariant under the action of the transfer matrix (exercise!).

For this reason one speaks of infinitely heavy external charges. It is doubtful whether there are covariant sectors corresponding to them in the continuum theory.

A particularly simple and interesting type of state are the "string states" : Select a path S_{xy} starting at x and ending at y in Λ_+ ; let $g_{S_{xy}} = P \prod_{<x',y'>\in S_{xy}} g_{x'y'}$ (P stands for path ordering).

Then $U_\tau(g_{S_{xy}})_{ab}$ will correspond to a state $\psi^{ab}_{\tau_x\bar{\tau}_y}(S_{xy})$ having charge τ at x and $\bar{\tau}$ at y.

Now a little computation shows that

$$(\psi^{ab}_{\tau_x\bar{\tau}_y}, T^t \psi^{cd}_{\tau_x\bar{\tau}_y}) \equiv \langle W_\tau(C_{xy,t})\rangle \delta_{ac}\delta_{bd} d_\tau^{-2} ,$$

where $W_\tau(C_{xy,t})$ is a Wilson loop observable, i.e.

$$W_\tau(C_{xy,t}) = \chi_\tau(P \prod_{<xy>\in C_{xy,t}} g_{xy})$$

and $C_{xy,t}$ is a closed path obtained by running first through S_{xy}, then straight up in time, then running backward through a shifted mirror image of S_{xy} and straight down in time to the starting point.

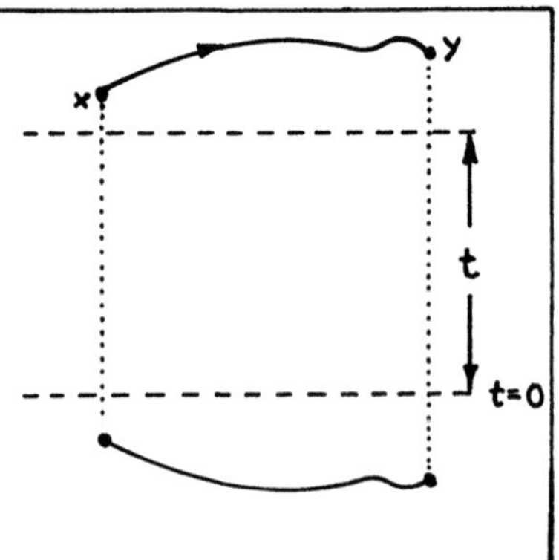

The loop $C_{xy,t}$

Starting with a straight "horizontal" string S_{xy} we obtain a rectangular Wilson loop.

If we define a Hamiltonian H by $T = e^{-2H}Q$ where Q is the projection onto the complement of the null space of T, we can express the "potential" between static charges as follows :

Let

$$V_{\tau_x\bar{\tau}_y}(S_{xy}) \equiv \inf \operatorname{spec} H \upharpoonright \psi^{ab}_{\tau_x\bar{\tau}_y}(S_{xy})$$

i.e. the lowest energy present in $\psi^{ab}_{\tau_x\bar{\tau}_y}(S_{xy})$. Then

$$V_{\tau_x\bar{\tau}_y}(S_{xy}) = - \lim_{t\to\infty} \frac{1}{2t} \log W_\tau(C_{xy,t})$$

(S_{xy} is assumed to be the shortest path from x to y).

If the Wilson loop now has "area decay", i.e.

$$\langle W_\tau(C)\rangle \leq \text{const} \exp(-\alpha_\tau A(C)) \quad , \text{ where } \alpha_\tau > 0$$

and $A(C)$ is the number of plaquettes in the minimal surface bordered by C , we obtain $V_{\tau_x\bar{\tau}_y}(S_{xy}) \geq \alpha_\tau \operatorname{dist}(x,y)$, i.e. the "potential" grows linearly with the distance. This gives some motivation for Wilson's confinement criterion.

Next we want to define a disorder operator analogous to the one introduced by 't Hooft in a slightly different framework. Let T be a set of (oriented) links in Λ_+ , ω an element of the center Z of G ; then we define a linear operator $\mathring{B}_\omega(T)$ on A_+ by mapping a function $f(\{g_{xy}\})$ into $f(\{\omega_{xy}(T)g_{xy}\})$ where $\omega_{xy}(T) = \omega$ if T contains $\langle xy\rangle$ with the right orientation and $\omega_{xy}(T) = \omega^{-1}$ if T contains $\langle yx\rangle$ with the right orientation.

Locally, for instance for single link, this amounts to a gauge transformation. In general, however, there will be plaquettes that are affected; the operation is therefore sometimes called a "singular gauge transformation". We denote the set of plaquettes in Λ_+ where the effects of $\mathring{B}_\omega(T)$ do not cancel by T' (it will be a subset of the "coboundary" δT of T , cf. [48]). Then we define

$$B_\omega(T') \equiv \exp\{\frac{1}{2g_o^2} \sum_{P\in T'} \chi(g_{\partial p})(\frac{\chi(\omega_{\partial p})}{\chi(\mathbb{1})} - 1)\}\, \mathring{B}_\omega(T)$$

$$= \mathring{B}_\omega(T)\exp\{\frac{1}{2g_o^2} \sum_{P\in T'} \chi(g_{\partial p})(1 - \frac{\chi(\omega^{-1}_{\partial p})}{\chi(\mathbb{1})})\}$$

where $\omega_{\partial P} = \omega^{n_+(\partial P\cap T)-n_-(\partial P\cap T)}$, $n_+(\partial P\cap T)$ $(n_-(\partial P\cap T))$ is the number of links of T occurring in ∂P with the right (wrong) orientation . $B_\omega(T')$ depends really only on T' ; it is defined in such a way that it preserves the physical scalar product and therefore gives rise to a unitary operator on the physical Hilbert space which we denote by the same symbol $B_\omega(T')$. (To see this, use the uncoupled expectation $\langle\cdot\rangle_o$) .

A computation shows that

$$B_{\omega_1}(T')B_{\omega_2}(T') = B_{\omega_1\omega_2}(T')$$

so that these operators form a unitary representation of the center Z .

Typically in d = 3 , T' will be chosen to be the set of all plaquettes dual (= orthogonal) to a string S_{xy} running from a point x to a point y , both situated in the plane t = 0 and in the dual lattice. An example is shown in the picture:

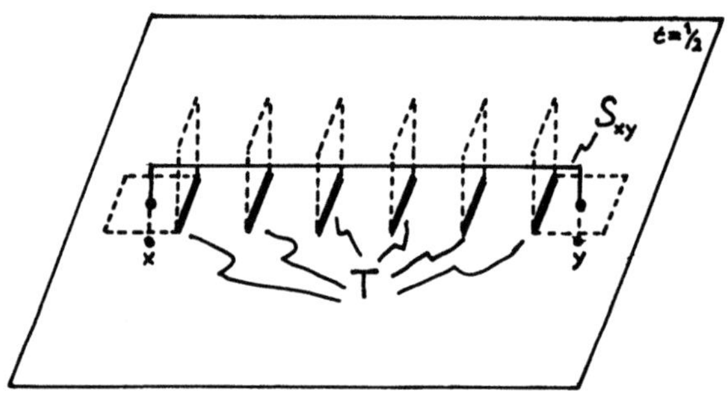

The indicated links form T , the indicated plaquettes T' . To denote the duality relationship between T' and S_{xy} we write $T' = *S_{xy}$. The endpoints x and y can be interpreted as "vortices". T can be described as dual to a sheet that is bordered by S_{xy} and a line in the t = 0 plane.

In d = 4 , T' will typically consist of all plaquettes dual to a surface S_C in the dual lattice intersecting the t = 0 plane in a closed loop C (a "vortex line"). T will be dual to a volume bordered by S_C and a surface in the t = 0 plane.

't Hooft [26] found interesting topological commutation relations between Wilson loops and disorder operators that help to interprete them in terms of each other:

In d = 3 consider a string *T' starting and ending in the t = 0 plane. T will then be dual to a sheet in Λ_+ bordered by *T' . Let $W_\tau(C)$ be a Wilson loop in Λ_+ . Then

$$B_\omega(T')W_\tau(C) = \chi_\tau(\omega^{n_{TC}})W_\tau(C)B_\omega(T') \quad ,$$

where n_{TC} is the oriented intersection number of C with T, or equivalently the winding number of C around $*T' \cup (*T \cap \{t=0\})$.

In $d = 4$ consider a surface $*T' \subset \Lambda_+$ starting with a closed loop C' in the $t = 0$ hyperplane and strectching upward in "time" direction. T will be dual to a 3-dimensional volume in Λ_+ bordered by $*T'$. Let $W_\tau(C)$ again be a Wilson loop in Λ_+ . Then again we have the commutation relations given above; n_{TC} is again the oriented intersection number of C with T but can also be interpreted as the linking number of C with $*T' \cup (*T \cap \{t=0\})$.

Since $W_\tau(C)$ in a plane t = const. can be said to measure magnetic flux, the commutation relations state that $B_\omega(T')$ is a creation operator for magnetic flux. Similarly one may say that $W_\tau(C)$ creates electric flux along C and $B_\omega(T')$ measures it.

Related but slightly different is the Wegner-'t Hooft disorder observable. It can be introduced as the dual of the Wilson loop in abelian models [5]. For that reason in $d = 4$ it is interpreted to be related to the potential between monopoles and antimonopoles in the same way as the Wilson loop is related to the potential between charges. A direct definition in $d = 4$ is as follows :

Let C be a closed loop in the dual lattice, S a sheet with $\partial S = C$. A modified partition function $Z_\Lambda(\omega, C)$ is defined by replacing $g_{\partial P}$ by $g_{\partial P}\omega$ on each plaquette in $*S$, the dual of S . Gauge invariance shows that actually only C , not S , matters. The 't Hooft loop expectation is then given by

$$\langle D_\omega(S)\rangle_\Lambda = \frac{Z_\Lambda(\omega,S)}{Z_\Lambda(\mathbb{1},S)}$$

We get the same result if we interprete $D_\omega(S)$ as an ordinary observable, namely (cf. [5]) :

$$D_\omega(S) = \exp\{\frac{1}{2g_o^2} \sum_{P\in *S} \chi(g_{\partial P}) \left(\frac{\chi(\omega)}{\chi(\mathbb{1})} - 1\right)\} \quad .$$

The monopole - antimonopole interpretation can also be carried through in an analogous way as the charge-pair interpretation for the Wilson loop : Taking a rectangular loop in the 01 plane and slicing in the t = const. hyperplane that contains just a horizontal piece we discover a picture like this :

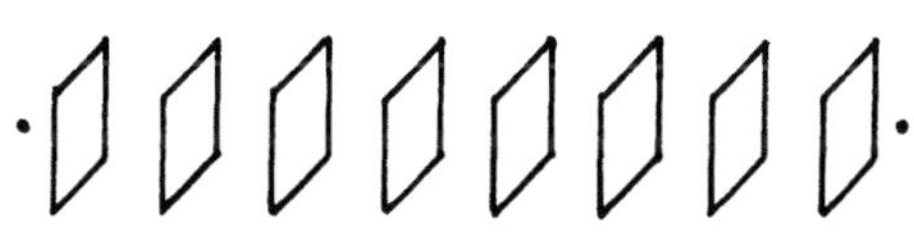

where the drawn plaquettes are all modified by ω . This means that the "Bianchi identity" is violated at the endpoints, i.e. the magnetic flux through the surface of a cube enclosing one of the endpoints is no longer zero (this interpretation is somewhat metaphoric for a non-abelian gauge theory). Therefore one identifies the endpoints as positions of a monopole-antimonopole pair.

c) "Diamagnetic" inequality.

The diamagnetic properties of bosonic systems are well known, at least for nonrelativistic systems. General proofs have been given by Simon [27, 29] and Hess, Schrader and Uhlenbrock [28], based on Kato's inequality or Itô's stochastic integrals, respectively (the latter method goes back to a remark by Nelson as quoted in [29]). For an overview of these and related facts see Hunziker [30].

Here we are dealing with essentially relativistic systems that may also contain fermions (spinors). The inequality to be proven below expresses a joint effect of the diamagnetic behavior of bosons and the paramagnetism due to the spin, hence the quotation marks. It is truly remarkable that these two opposite effects produce inequalities of the same kind in this field theoretic context.

The "diamagnetic" inequalities deal with partition functions in an external gauge field : Let $\langle\cdot\rangle_{o,\Lambda,\{g_{xy}\}}$ be the "conditional matter expectation" (no integration over $\{g_{xy}\}$) ; then

$$Z_\Lambda(\{g_{xy}\}) \equiv \langle e^{-\overset{\circ}{S}_H - S_F}\rangle_{o,\Lambda,\{g_{xy}\}} \tag{2.8}$$

The inequality says simply

$$|Z_\Lambda(\{g_{xy}\})| \leq Z_\Lambda(\{\mathbf{1}\}) \tag{2.9}$$

There are proofs of varying generality depending on whether G is abelian or nonabelian and whether fermions are present or not (see [23,31]).

The most "natural" proof of (2.9) would go through showing that the left hand side is of positive type in $\{g_{xy}\}$ or at least expressible as a polynomial in Wilson loops with positive coefficients. Unfortunately this does not seem to be true in the presence of fermions. So in the most general framework we only have a rather restricted result :

Theorem 2.3. : Let Λ be a hypercube with (anti-) periodic boundary conditions. Then the "diamagnetic" bound (2.9) holds.

Proof : Let $C \equiv \sup_{\{g_{xy}\}} |Z_\Lambda(\{g_{xy}\})|$.

Since the space of $\{g_{xy}\}$ is compact and Z_Λ is a continuous function on it we may assume

$$C = |Z_\Lambda(\{\mathring{g}_{xy}\})| \quad . \tag{2.10}$$

Let π be any pair of antipodal hyperplanes lying midway between "fixed time" lattice hyperplanes. As discussed in the proof of Theorem 2.1, we may choose a gauge such that $\{\mathring{g}_{xy}\}$ does not contain any group elements different from **1** on the links crossing π . This produces

$$e^{-\mathring{S}_H - S_F} = e^{-S_+}\theta_M(e^{-\tilde{S}_+})e^{-S_C}$$

with S_C of the form

$$-S_C = \sum_i G_i \theta_M G_i' \quad , \quad G_i \in A_{\Lambda_+}$$

(here θ_M is defined like θ but does not operate on the gauge field).

Now the expectation $\langle\cdot\rangle_{o,\Lambda,\{g_{xy}\}}$ is trivially O.S. positive with respect to θ_M and we can use the following

Lemma 2.4 : If $\langle\cdot\rangle$ is O.S. positive, $F, G_i \in A_{\Lambda_+}$ (for all i) , then :

$$|\langle F\theta_M F' e^{\Sigma G_i \theta_M G_i'}\rangle| \leq$$

$$\leq \langle F\theta_M F\, e^{\Sigma G_i \theta_M G_i}\rangle^{1/2} \langle F'\theta_M F' e^{\Sigma G'_i \theta_M G_i'}\rangle^{1/2}$$

Proof : The left hand side is

$$\Big|\sum_{\substack{i_1,\dots,i_n\\ n}} \frac{1}{n!} \langle F\theta_M F' G_{i_1}\dots G_{i_n}\theta_M(G_{i_1}'\dots G_{i_n}')\rangle\Big| \leq$$

$$\leq \sum_{\substack{i_1,\dots,i_n\\ n}} \frac{1}{n!} \langle F\theta_M F\, G_{i_1}\dots G_{i_n}\theta_M(G_{i_1}\dots G_{i_n})\rangle^{1/2}$$

$$\times \langle F'\theta_M F'\, G_{i_1}'\dots G_{i_n}'\, \theta_M(G_{i_1}'\dots G_{i_n}')\rangle^{1/2}$$

$$\leq \Big(\sum_{i_1,\dots,i_n} \frac{1}{n!} \langle F'\theta_M F' G_{i_1}'\dots G_{i_m}' \theta_M(G_{i_1}'\dots G_{i_m}')\rangle\Big)^{1/2} \times$$

$$\times \Big(\sum_{j_1,\dots,j_m} \frac{1}{m!} \langle F'\theta_M F' G'_{j_1} \dots G'_{j_m} \theta_M (G'_{j_1} \dots G'_{j_m}) \rangle \Big)^{1/2}$$

which equals the right hand side.

□

Cor. 2.5 : Let $g_{xy} = \mathbf{1}$ for xy crossing π . Then

$$|Z_\Lambda(\{g_{xy}\})| \le \langle e^{-S_+} \theta_M e^{-S_+} e^{G_i \theta_M G_i} \rangle^{1/2} \times$$

$$\times \langle \theta_M e^{-S_-} e^{-S_-} e^{\Sigma(\theta_M G') G'_i} \rangle^{1/2} \equiv$$

$$\equiv Z_\Lambda(\{g^+_{xy} \theta g^+_{xy}\})^{1/2} \, Z(\{(\theta g^-_{xy}) g^-_{xy}\})^{1/2} \quad .$$

Here $g^\pm_{xy}$ stands for the part of g_{xy} living in $\Lambda^\pm$, $g^+_{xy}\,\theta g^+_{xy}$, etc. is the gauge field consisting of g^+_{xy} in Λ^+ and its reflection in Λ^- .

Proof : This is a direct application of the lemma.

□

Now observe that $g^+_{xy}\,\theta g^+_{xy}$ is trivial (gauge equivalent to $\mathbf{1}$) on all plaquettes bisected by π .

Using (2.10) we obtain

$$|Z_\Lambda(\{\mathring{g}_{xy}\})| \le Z_\Lambda(\{\mathring{g}^+_{xy}\;\theta \mathring{g}^+_{xy}\}) \tag{2.11}$$

Assume now that $\mathring{g}_{xy}$ is already trivial on N plaquettes. If we manage to choose our pair of hyperplanes π in such a way that Λ_+ contains more than $\frac{N}{2}$ of those trivial plaquettes $\mathring{g}^+_{xy}\;\theta\mathring{g}^+_{xy}$ will contain at least N+1 trivial plaquettes. If that is not possible, a little thought shows that applying (2.11) once will make the distribution of trivial plaquettes unsymmetrical so that in the next step N may be increased.

Applying this argument a finite number of times we see that $Z_\Lambda(\{\mathring{g}_{xy}\})$ can be bounded by $Z_\Lambda(\{\mathring{\mathring{g}}_{xy}\})$ where $\{\mathring{\mathring{g}}_{xy}\}$ has only trivial plaquettes. Because our periodic lattice contains nontrivial (noncontractible) loops this does not yet quite imply that $\mathring{\mathring{g}}_{xy}$ is gauge equivalent to $\mathbf{1}$. But by reflecting about hyperplanes in all d directions of the lattice and using (2.11) we will obtain a completely trivial gauge field. This finishes the proof of Theorem 2.3.

□

If we are dealing with purely bosonic matter and if the representation U_H fulfils a certain condition (R) (see below), a more general result holds :

Theorem 2.6 : Let Λ be any finite lattice (not necessarily part of a regular lattice). If U_H obeys the condition (R) , (2.9) holds.

Remark : A finite lattice is nothing but a finite graph, i.e. a collection of sites (vertices) and links (lines) with an incidence relation between them.

Next we define the condition (R).

Definition 2.7. Let N_o be the smallest integer - if it exists - such that $U_H^{\otimes_s N_o}$ (the N_o fold symmetric tensor product) is the trivial representation, otherwise we set $N_o = \infty$. We say that U_H has the property (R) iff either U_H is a real representation and for any $\phi \in U_H$ with $||\phi|| = 1$

$$\int dh \prod_{i=1}^{N} (U_H(h)\phi)_{a_i} = \begin{cases} 0 & \text{if } N \text{ is odd} \\ \text{const} \sum_{\substack{\text{pairings} \\ \{i,p(i)\}}} \Pi\delta_{a_i a_{p(i)}} & \text{for } N \text{ even} \end{cases} \tag{2.12}$$

or U_H is a complex representation and for any $\phi \in U_H$ with $||\phi|| = 1$ and for $N,M < N_o$

$$\int dh \prod_{i=1}^{N} (U_H(h)\phi)_{a_i} \prod_{j=1}^{M} (U_H(h)\phi)_{b_j} = \delta_{NM} \sum_{\substack{\text{permutations} \\ \pi \in S_n}} \frac{1}{N!} \prod_{i=1}^{N} \delta_{a_i b_{\pi(i)}} \tag{2.13}$$

(We chose the name because property (R) allows to "route" representations and Wilson loops through vertices). Let us give two classes of examples for property (R) :

Lemma 2.8 : (1) If G is abelian, any irreducible representation has the property (R) .

(2) If $G = SU(N)$, the fundamental representation has the property (R).

Proof of Lemma 2.8 : (1) We may restrict ourselves to the groups $\mathbb{Z}_n$ and $U(1)$ since everything else can be obtained by forming direct products of these.

a) $\mathbb{Z}_n$: Let the representation U_H be given by

$$U_H(e^{i\frac{2\pi}{n}}) = e^{\frac{2\pi i}{n}k} .$$

It is real if $n = 2k$; in this case property (R) is trivial. If $n \neq 2k$ (2.13) can be easily checked.

b) U(1) : It suffices to consider complex representations; (2.13) is trivial. □

Proof of (2) : The fundamental representation of SU(N) is complex. Let

$$V_{\substack{a_1,\dots,a_N\\ b_1,\dots,b_M}} \equiv \int dh \prod_{i=1}^{N} (U_H(h)\phi)_{a_i} \prod_{j=1}^{M} \overline{(U_H(h)\phi)}_{b_j}$$

Obviously this can be interpreted as a map from the symmetric subspace of $\bigotimes_{j=1}^{M} V_H$ to the symmetric subspace of $\bigotimes_{j=1}^{N} V_H$.

On each of these spaces an irreducible representation U_N^S or U_M^S respectively, is operating and

$$U_N^S(g)V = VU_M^S(g) .$$

This implies already that $V = 0$ unless $M = N$ (Schur's lemma). But for $M = N$ we see that V commutes with U_N^S and is therefore a multiple of the identity. The multiplying factor can be determined by taking the trace. □

We leave it as an exercise to show that the fundamental representations of O(N) have the property (R) in the form (2.12). One has to use the decomposition of the symmetric tensor product into irreducible representations and Schur's lemma.

Proof of Theorem 2.6 :

$$Z_\Lambda(\{g_{xy}\}) = \int e^{-\overset{\circ}{S}_H} \prod_{x\in\Lambda} d\rho(|\phi(x)|) \tag{2.14}$$

where $d\rho(|\phi|) = e^{-V(|\phi|)}d\phi$.

If we expand the exponential we obtain

$$Z_\Lambda(\{g_{xy}\}) = \sum_{\{n_{xy}\}} \prod_{\langle xy\rangle} \frac{1}{n_{xy}!} \int (\phi(x),U_H(g_{xy})\phi(y))^{n_{xy}} \times \\ \times \prod_x d\rho(|\phi(x)|) . \tag{2.15}$$

We claim

Lemma 2.9 : If U_H has the property (R)

$$\prod_{<xy>} \int (\phi(x), U_H(g_{xy})\phi(y))^{n_{xy}} \prod_x d\rho(|\phi(x)|) =$$

$$= \sum_L c_L \prod_{C \in L} \chi_H(P \prod_{<xy> \in C} g_{xy}) \tag{2.16}$$

where $c_L \geq 0$ and P is a path ordering symbol. L labels systems of closed loops $\{C_1, \ldots, C_{|L|}\}$ in Λ ; χ_H stands for the character belonging to the representation U_H .

<u>Corollary 2.9'</u> : If G is abelian, $Z(\{g_{xy}\})$ is of positive type in $\{g_{xy}\}$.

We note that Lemma 2.9 and eq. (2.15) immediately give the diamagnetic bound (2.9).

<u>Proof of Lemma 2.9</u> : Consider the following expression arising as a generic factor of the left hand side of eq. (2.16) :

$$\int \prod_{i=1}^{N} (\phi(x_i), U_H(g_{x_i x})\phi(x)) \times$$
$$\times \prod_{j=1}^{M} (\phi(x), U_H(g_{xy_j})\phi(y_j)) d\rho(|\phi(x)|) \quad .$$

because of the invariance of the measure $d\rho$ this can be rewritten as

$$\int dh \int \prod_{i=1}^{N} (\phi(x_i), U_H(g_{x_i x}) U_H(h)\phi(x)) \times$$
$$\times \prod_{j=1}^{M} (U_H(h)\phi(x), U_H(g_{xy_j})\phi(y_j)) d\rho(|\phi(x)|) \quad . \tag{2.17}$$

By property (R) either (2.12) or (2.13) holds; in the first case the expression (2.17) can be written as

$$\text{const.} \sum_{\substack{\text{pairings} \\ \{i,p(i)\}}} \prod_{i=1}^{N+M} (\phi(x_i), U_H(g_{x_i x} g_{x x_{p(i)}})\phi(x_{p(i)})) \times$$
$$\times \int d\rho(t) t^{N+M} \tag{2.18}$$

(if $N+M$ is even, otherwise we obtain 0), where we put

$$x_{N+j} \equiv y_j \quad (j = 1, \ldots, M) \quad .$$

In the second case (2.17) becomes for $N_o = \infty$

$$\delta_{NM} \sum_{\pi \in S_N} \frac{1}{N!} (\phi(x_i), U_H(g_{x_i x} g_{xy_{\pi(i)}})\phi(y_{\pi(i)})) \times \int d\rho(t) t^{2N} \quad . \tag{2.19}$$

Inserting (2.18) or (2.19), respectively into the left hand side of (2.16) we obtain an expression of the form asserted in Lemma (2.16). Actually the coefficients c_L could be computed explicitly, but in any case they are clearly nonnegative. The modifications necessary for $N_o < \infty$ are straightforward. □

Remark : For fermionic matter it is not clear whether anything as general as Theorem 2.6 is true. In particular the analogue of Lemma 2.9 does not seem to hold. There are actually simple closed loops that give a negative contribution [32]. The "diamagnetic" bound for spinors seems to be of a much subtler nature; it depends on the fact that "on the average" the paramagnetic effect of the spin dominates the diamagnetism that is also present. Note in this connection that the nonrelativistic "paramagnetic conjecture" [33] was shown to fail for Aharonov-Bohm like fields [34].

d) Correlation Inequalities.

Correlation inequalities of a sufficiently general type have only been found for abelian gauge theories with purely bosonic matter.

For Wilson's or Villain's form of the action they follow rather directly from Ginibre's general results (see [35,36]). In [23] analogous inequalities were proven for a Gaussian action for the gauge field. The inequalities are of the general form

$$\langle AB\rangle_\Lambda - \langle A\rangle_\Lambda \langle B\rangle_\Lambda \equiv \langle A;B\rangle_\Lambda \geq 0$$

where A,B belong to the multiplicative cone generated by

$$\{|\phi(x)| \; | x \in \Lambda\}$$

and

$$\{\cos(\sum_x m_x \theta_x + \sum_{\langle xy\rangle} f_{xy} A_{xy})\}$$

where we defined $\theta_x \equiv \arg \phi(x)$, $A_{xy} \equiv \arg g_{xy}$, m_x is integer valued; f_{xy} is integer valued if we are dealing with the Wilson or Villain actions, real valued for the Gaussian action.

We summarize some consequences of these and some related inequalities that are of importance for the continuum Higgs_2 model to be discussed in the second part of these lectures :

Theorem 2.10 : Let $\langle\cdot\rangle_{C,(\Lambda)}$ be the expectation of an abelian Higgs model where the measure for the gauge field A is Gaussian with covariance C .

Then

(1) $\quad <e^{|\phi|^2(g)} e^{iA(f)}>_C \quad$ is decreasing in C .

(2) $\quad <e^{-|\phi|^2(g)} e^{A(f)}>_C \quad$ is increasing in C .

(3) $\quad <e^{|\phi|^2(g)} e^{iA(f)}>_{C,\Lambda} \quad$ is increasing in Λ .

(4) $\quad <e^{-|\phi|^2(g)} e^{A(f)}>_{C,\Lambda} \quad$ is decreasing in Λ .

where $|\phi|^2(g) = \sum_x |\phi(x)|^2 g(x) \quad , \quad g \geq 0 \quad , \quad A_{xy} = \arg g_{xy}$.

For details of the proof which is elementary but lengthy and patterned after Ginibre's proof [35] we refer to [23].

We remark that these correlation inequalities can be used to deduce some monotonicity properties of the potential between "infinitely heavy quarks" of opposite charge [31]. But since this refers to an abelian theory it is not all that relevant for the confinement problem.

We close this section by mentioning two problems that are worthy of further investigation (exercises, if you like) :

(1) Give a more general proof of the "diamagnetic" inequality, both for Higgs matter in an arbitrary representation and for fermions.

(2) Find sufficiently strong correlation inequalities for non-abelian groups.

3. EXPANSION METHODS.

The method of high and low temperature cluster expansions is one of the best known techniques in statistical mechanics. Knowing that I will be bringing coal to Newcastle I will nevertheless give a general and self-contained exposition with some personal twists in it and then focus on applications in lattice gauge theories.

Gruber and Kunz [37] showed that a general class of lattice systems can be mapped into a polymer system of the kind they studied in detail. That this was a useful point of view for lattice gauge theories was pointed out by Gallavotti et al [38] (this short paper is, however, very sketchy in the discussion of the combinatoric aspects). How to map lattice systems into polymer systems can be learned from the very neat paper by Gallavotti, Martin-Löf and Miracle-Solé on the low temperature expansion of the Ising model [39]. This is the approach we will follow here. The treatment of the combinatorics is very much inspired by Malyshev's work [42].

The idea is the following : In the "high temperature region", that is the region of parameters where the coupling between different sites or different links, respectively, is weak, it is reasonable to write

$$e^{\frac{1}{2g_o^2}\chi(g_{\partial P})} = 1 + \rho_{\partial P}$$

$$e^{\frac{\kappa}{2}\bar{\psi}(x)U\Gamma\psi(y)} = 1 + \lambda_{<xy>}$$

$$e^{\frac{\lambda}{2}(\phi(x),U(g_{xy})\phi(y))} = 1 + \mu_{<xy>}$$

and expand the expectation value of an observable A in the small parameters $\rho_{\partial P}$, $\lambda_{<xy>}$, $\mu_{<xy>}$ (for the Higgs action we will see that there is a modification possible that gives a much wider range of convergence - this is the so-called Higgs mechanism in disguise). Thereby we obtain decoupled expectations of the form

$$<A \prod_{P\in\gamma_P} \rho_{\partial P} \prod_{<xy>\in\gamma_H} \lambda_{<xy>} \prod_{<xy>\in\gamma_F} \mu_{<xy>}>_{0,\Lambda} \tag{3.1}$$

where γ_P is a set of plaquettes, γ_H a set of Higgs links, γ_F a set of Fermion links. We call $\gamma_P \cup \gamma_H \cup \gamma_F$ a set of <u>bonds</u>. Now the expectation $<AB>_{0,\Lambda}$ factorizes into $<A>_{0,\Lambda} <B>_{0,\Lambda}$ whenever A and B depend on disjoint sets of variables. We call a set γ of bonds that does not break up into subsets corresponding to disjoint sets of field variables a <u>polymer</u> (roughly this means that γ has to be connected geometrically, but for instance two plaquettes touching at a corner are not considered connected). Of course we have to take the variables occurring in A into account, so a polymer γ occurring in (3.1) may be connected just through the "support" of A. We call

$$z_A(\gamma) \equiv <A \prod_{P\in\gamma} \rho_{\partial P} \prod_{<xy>\in\gamma} \lambda_{<xy>} \prod_{<xy>\in\gamma} \mu_{<xy>}>$$

the <u>activity</u> of the polymer γ . So we obtain an expansion in activities of polymers for both $<A \exp(-\mathring{S}_H-S_F-S_{Y.M.})>_{0,\Lambda}$ and for $Z_\Lambda = <\exp(-\mathring{S}_H-S_F-S_{Y.M.})>_{0,\Lambda}$.

The art of the game is

(1) to use the techniques of formal power series that were first brought into this context by Ruelle [40] in order to form the quotient expressing the full expectation

$$\langle A\rangle_\Lambda = \frac{1}{Z_\Lambda} \langle A\, e^{-\mathring{S}_H - S_F - S_{Y.M.}} \rangle_{0,\Lambda}$$

and

(2) to show that the expansion has a finite domain of convergence independent of the volume Λ . The thermodynamic limit can then be performed term by term which is easy. One obtains a unique thermodynamic limit (independent of boundary conditions) and also exponential clustering, "confinement" and other desirable results.

The same technique can also be applied to the "low temperature" region which is typically given by g_o^2 small, γ and κ small. But in this case the polymers will correspond to "defects", i.e. sets where $g_{\partial P} \neq \mathbb{1}$ (also called vortices) and the sets of Higgs and Fermi links connected to them in an appropriate way. It is clear that this can only work for discrete groups G ; for $G = \mathbb{Z}_2$ this was first carried out by Marra and Miracle-Solé [41]. The word "defect" is not chosen arbitrarily; in fact lattice gauge theories with discrete gauge group have been proposed as realistic models for defects in ordered media [7].

There are other types of expansions, such as the $1/d$ expansion (see for instance [82,83]) which will not be discussed here.

My treatment of the expansions is certainly not original (this would probably be impossible !), but it is not quite identical to any treatment I found in the literature. In particular I tried to keep separate as much as possible purely combinatorial facts and estimates depending on the special structure of the lattice (in particular its dimensionality), very much in the spirit of Malyshev [42].

a) General algebraic formalism for polymers

Here we develop the theory of cluster expansions for polymers abstractly, using the method of formal power series; the presentation is very much inspired by [39].

Let Γ_o be a (finite) set whose elements $\gamma_1, \gamma_2, \ldots$ are called polymers. We assume that a function

$$g : \Gamma_o \times \Gamma_o \to \{0,-1\}$$

is given such that $g(\gamma,\gamma) = -1$ for all $\gamma \in \Gamma_o$. We say that γ, γ' are compatible iff $g(\gamma,\gamma') = 0$ and incompatible otherwise.

For each $\gamma \in \Gamma_o$ there is an "activity" $z(\gamma)$ which is sometimes interpreted as an indeterminate and sometimes as a complex number; from the context it will be clear which interpretation is assumed.

Def. 3.1. :

$$Z(\{z(\gamma)\}_{\gamma\in\Gamma_o}) \equiv \sum_{\Gamma\in\Gamma_o} \prod_{\gamma\in\Gamma} z(\gamma) \prod_{\substack{i<j \\ \gamma_i,\gamma_j\in\Gamma}} (1+g(\gamma_i,\gamma_j)) \tag{3.2}$$

is called partition function.

$$\phi(\Gamma) \equiv \prod_{\gamma\in\Gamma} \prod_{\substack{i<j \\ \gamma_i,\gamma_j\in\Gamma}} (1+g(\gamma_i,\gamma_j)) \tag{3.3}$$

is called Boltzmann factor.

We will denote by X, Y etc. functions $\Gamma_o \to \mathbb{N}$, i.e. multi-indices and we define

$$z^X \equiv \prod_{\Gamma\in\Gamma_o} z(\gamma)^{X(\gamma)} \tag{3.4}$$

to be the corresponding monomial.

$$X! \equiv \prod_{\Gamma\in\Gamma_o} X(\gamma)! \quad ; \quad n(X) \equiv \sum_{\gamma\in\Gamma_o} X(\gamma) \ . \tag{3.5}$$

If $X! = 1$ we may identify X with the subset

$$\Gamma_X = \{\gamma \in \Gamma_o | X(\gamma) = 1\} \ .$$

If we define

$$\phi(X) \equiv \begin{cases} \phi(\Gamma_X) & \text{if } X! = 1 \\ 0 & \text{if } X! > 1 \end{cases}$$

we can write therefore

$$Z = \sum_X z^X \phi(X)$$

Def. 3.2 : Let f_1 , f_2 be functions from the set of multiindices into $\mathbb{C}$. Then

$$(f_1 * f_2)(X) \equiv \sum_{X_1+X_2=X} f_1(X_1) f_2(X_2) \tag{3.6}$$

Remark : If we interprete f_1 , f_2 as coefficient sequences of formal power series then this definition makes $f_1 * f_2$ the coefficient sequence of the product.

Def. 3.3 :
$$\phi^T(X) \equiv (\text{Log } \phi)(X)$$

is called "Ursell function", where for any function f ,

$$(\text{Log } f)(X) = \sum_{n\geq 1} \frac{(-1)^{n+1}}{n} (f-1)^{*n}(X) \ ;$$

$$\mathbb{1}(X) = \begin{cases} 1 & \text{if } X = 0 \\ 0 & \text{otherwise} \end{cases}$$

Prop. 3.4 :
$$\log Z = \sum_X \phi^T(X) z^X$$

Proof : This follows from the remark after Def. 3.2. □

We now prove an explicit formula for $\phi^T(X)$ given by Gallavotti, Miracle-Solé and Martin-Löf [39] and that is familiar from the theory of the Mayer expansion.

Lemma 3.5 :
$$\phi^T(X) = \frac{1}{X!} a(X) \qquad (3.7)$$

where

$$a(X) = \sum_{C \subset G(X)} (-1)^{\ell(C)} \qquad (3.8)$$

Here $G(X)$ is the graph with vertices $\gamma_1, \ldots, \gamma_{n(X)}$ (take $X(\gamma)$ copies of each γ) and a line connecting γ , γ' ($\gamma \neq \gamma'$) whenever $g(\gamma,\gamma') = -1$; the sum is over connected subgraphs C having the same vertices, $\ell(C)$ the number of lines in C .

Proof :

$$\phi(X) = \prod_{\substack{i<j \\ X(\gamma_i), X(\gamma_j) > 0}} (1+g(\gamma_i,\gamma_j)) \qquad (3.9)$$
$$= \sum_G \prod_{\{i,j\} \in G} g(\gamma_i,\gamma_j)$$

where the sum is over all graphs on $N = \{1,2,\ldots,n(X)\}$, i.e. all sets of two-element subsets of N (note that such graphs have at most one line between any two vertices). Let $(\gamma_1,\ldots,\gamma_{n(X)})$ be such that there are $X(\gamma)$ copies of each γ

in it. Define for $M \subset N$:

$$g(M) \equiv \begin{cases} \sum\limits_{C_M} \prod\limits_{\substack{i<j \\ i,j \in M}} g(\gamma_i,\gamma_j) & \text{if } |M| \geq 2 \\ 1 & \text{if } |M| = 1 \\ 0 & \text{if } M = \emptyset \end{cases} \tag{3.10}$$

($\sum\limits_{C_M}$ is over all connected graphs on M , i.e. with vertex set M). Then

$$\phi(X) = \sum\limits_{\sum_{i=1}^{m} N_i = N}' g(N_i)\cdots g(N_m) \tag{3.11}$$

(Σ' means sum over <u>different</u> partitions of N into $N_1,\ldots N_m$; for given $|N_1|,\ldots,|N_m|$ there are $\frac{1}{m!}\ \frac{n(X)!}{|N_1|!\cdots|N_m|!}$ of those).

We define also

$$g(\gamma_1,\ldots,\gamma_N) \equiv \begin{cases} \sum\limits_{C} \prod\limits_{\{i,j\}\in C} g(\gamma_i,\gamma_j) & \text{if } N \geq 2 \\ 1 & \text{if } N = 1 \\ 0 & \text{if } N = 0 \end{cases} \tag{3.12}$$

($\sum\limits_{C}$ is over all connected graphs on $\gamma_1,\ldots,\gamma_N$)

In order to figure out $(\mathrm{Log}\ \phi)(X) = \phi^T(X)$ it is useful to represent the formal power series for Z in a slightly different form :

$$Z = \sum_{n=o}^{\infty} \frac{1}{n!} \sum_{\gamma_1,\ldots,\gamma_n} \prod_{i=1}^{n} z(\gamma_i) \prod_{j<\ell} (1+g(\gamma_j,\gamma_\ell)) \tag{3.13}$$

(sum over <u>ordered</u> sequences $\gamma_1,\ldots,\gamma_n$) . Using (3.11) this becomes

$$Z = \sum_{n=o}^{\infty} \frac{1}{n!} \sum_{\gamma_1,\ldots,\gamma_n} \prod_{i=1}^{n} z(\gamma_i) \sum_{\sum N_i = n} g(N_1)\cdots g(N_m) \tag{3.14}$$

We now reorder this sum by first summing over γ's that make up the "clusters" corresponding to $N_1,\ldots,N_m$, keeping the sizes $n_1 = |N_1|,\ldots,n_m = |N_m|$ fixed. We obtain

$$Z = \sum_{n=o}^{\infty} \frac{1}{n!} \sum_{m=o}^{\infty} \sum_{\substack{n_1,\dots,n_m \\ \Sigma n_i = n}} \frac{1}{m!} \frac{n!}{n_1!\cdots n_m!}$$

$$\prod_{i=1}^{n} \Big(\sum_{\gamma_1,\dots\gamma_{n_i}} g(\gamma_1,\dots,\gamma_{n_i}) \prod_{j=1}^{n_i} z(\gamma_j) \tag{3.15}$$

or

$$Z = \sum_{m=o}^{\infty} \frac{1}{m!} \Big(\sum_{k=1}^{\infty} \frac{1}{k!} \sum_{\gamma_1,\dots,\gamma_k} g(\gamma_1,\dots,\gamma_k) \prod_{j=1}^{k} z(\gamma_j)\Big)^m \tag{3.16}$$

This means that

$$\log Z = \sum_{k=1}^{\infty} \frac{1}{k!} \sum_{\gamma_1,\dots,\gamma_k} g(\gamma_1,\dots,\gamma_k) \prod_{j=1}^{k} z(\gamma_j) \tag{3.17}$$

Comparing coefficients with $\log Z = \Sigma\, z^X \phi^T(X)$ we obtain

$$\phi^T(X) = \frac{1}{X!}\, g(\gamma_1,\dots,\gamma_{n(X)}) \tag{3.18}$$

where $(\gamma_1,\dots,\gamma_{n(X)})$ contains exactly $X(\gamma)$ copies of each γ . Looking at the definition (3.12) this gives (3.7) and (3.8).

□

Now we have to estimate the graph theoretic expression

$$a(G) \equiv \sum_{C\subset G} (-1)^{\ell(C)} \tag{3.19}$$

(sum over connected subgraphs with the same vertex set V as G) . This is the content of

Theorem 3.6 : $|a(G)| \leq |T(G)|$ where $T(G)$ is the set of trees of G (i.e. trees with the same vertex set V as G and contained in G).

Remark : The proof of this theorem is in its essence due to Rota [43] ; I learned it through Malyshev's work [42] . In both references it is, however, buried in the rather heavy formalism of Möbius functions which we will not need here. A proof can also be found in Penrose's work [43].

Proof : We order the lines of the graph G any way we want and keep this order fixed. This induces an ordering of the loops (= elementary 1-cycles) of the graph : A loop L can be considered as an ordered sequence of lines : $L = (\ell_{i_1},\dots,\ell_{i_k})$ such that $i_1 < i_2 < \dots < i_k$. We then say that $\tilde{L} \geq L$ if the last line of $\tilde{L}$ has an index

$\tilde{i}_k \geq i_k$. This can be made into a complete ordering of the loops $L_1,\ldots,L_\sigma$ by proceeding lexicographically (backward).

Def. 3.7 : A broken loop is a loop minus its highest line.

Note that a loop L can be unambiguously reconstructed from the corresponding broken loop L' since there is never more than one line between two vertices. We order the broken loops $L_1',\ldots,L_\sigma'$ according to the order chosen for the loops.

The idea of the proof if now to reduce the sum over connected graphs (3.19) by successively excluding broken loops until there are only trees left.

Without loss of generality assume $|V| \geq 3$. Let

$R_j^i \equiv$ set of connected subgraphs of G of j lines, not containing $L_1',\ldots,L_i'$ (i.e. not containing the lines of $L_1',\ldots,L_i'$)

$R_j^{i,A} \equiv$ set of connected subgraphs of G of j lines, not containing $L_1',\ldots,L_{i-1}'$, but containing L_i

$R_j^{i,B} \equiv$ set of connected subgraphs of G of j lines, not containing $L_1',\ldots,L_{i-1}'$, but containing L_i' , not however L_i .

(A connected subgraph of G is always understood to have the same vertex set as G).

Clearly this gives a decomposition of R_j^{i-1} :

$$|R_j^{i-1}| = |R_j^i| + |R_j^{i,A}| + |R_j^{i,B}| \tag{3.20}$$

We now claim :

For $i = 0,1,\ldots$

$$a(G) = \sum_{j\geq 2} (-1)^j |R_j^i| \tag{3.21}$$

This is true for $i = 0$ by definition. Assume it is proven for $i-1$. Then by (3.20)

$$a(G) = \sum_{j\geq 2} (-1)^j |R_j^i| + |R_2^{i,A}| + \sum_{j\geq 2} (-1)^j (|R_j^{i,B}| - |R_{j+1}^{i,A}|) \tag{3.22}$$

Now $R_2^{i,A} = \emptyset$ because there are no loops of two lines; furthermore a bijection from $R_j^{i,B}$ to $R_{j+1}^{i,A}$ is easily constructed by adding the line $L_i \setminus L_i'$ (notice that this cannot violate any one of the conditions concerning $L_1',\ldots,L_{i-1}'$ since the added line has a higher index from any line in $L'_1,\dots,L'_{i-1})$.

So (3.21) is established. We use it now for $i = \sigma$: R^{σ}_{j} consists only of trees; hence all its elements have $|V| -1$ lines and we obtain

$$a(G) = (-1)^{|V|-1}\left|R^{\sigma}_{|V|-1}\right| \qquad (3.23)$$

from which Theorem 3.6 follows. □

Let now $C^G(\gamma)$ be the number of lines of G incident with the vertex γ . Then we have

Theorem 3.8 : $|T(G)| \leq \prod_{\gamma\in V} C^G(\gamma)$.

Proof : It suffices to construct for each tree T a map ψ_T from the vertex set V into the set $L(T) = \{\text{lines of } T\}$ such that $\psi_T(\gamma)$ is incident with γ and $\psi_T \neq \psi_{T'}$ for $T \neq T'$.

This is accomplished most easily by first defining an injective map φ_T : $L(T) \to V$ as follows :

We order the lines of T such that ℓ_{m+1} shares one vertex with at least one of the lines $\ell_1,\dots,\ell_m$, whereas its other vertex is not contained among the endpoints of $\ell_1,\dots,\ell_m$. φ_T is then defined inductively : $\varphi_T(\ell_1)$ is any endpoint of ℓ_1 ; $\varphi_T(\ell_{m+1})$ is the endpoint of ℓ_{m+1} shared with another line in $\{\ell_1,\dots,\ell_m\}$, provided that it did not yet occur among $\varphi_T(\ell_1),\dots,\varphi_T(\ell_m)$; if that is the case we take the other endpoint.

Now we can define

$$\psi_T(\gamma) = \begin{cases} \varphi_T^{-1}(\gamma) & \text{if defined} \\ \text{any line of } T \text{ incident with } \gamma & \text{otherwise} \end{cases}$$

The image $\psi_T(V)$ clearly consists of all lines of T , hence different trees give rise to different maps ψ_T and the theorem follows.

We have thus obtained the following estimate for $a(X)$ (eq. (3.8)) :

Cor. 3.9 : Let $C^X(\gamma) = -\sum_{\gamma'} X(\gamma')g(\gamma,\gamma')-1$ (3.24)

i.e. the number of lines incident with the vertex γ in the graph $G(X)$ corresponding to X . Then

$$|a(X)| \le \prod_{\gamma} C^X(\gamma)^{X(\gamma)} \tag{3.25}$$

Proof : Obvious. □

Remarks : (1) To check how sharp Cor. 3.9 is we may look at a system containing exactly one polymer γ with activity z . This has partition function $Z = 1+z$,

$$\log Z = \sum_{n=1}^{\infty} (-1)^{n-1} \frac{1}{n} z^n = \Sigma \frac{1}{X!} a(X) z^X$$

from which one reads off

$$a(X) = (n(X)-1)!(-1)^{n(X)-1}$$

The corresponding graphs are so-called complete graphs (all possible line drawn) and Cor. 3.9 gives

$$|a(X)| \le (n(X)-1)^{n(X)}$$

which is too large by about a factor $e^{n(X)}$.

(2) Cor. 3.9 is the best estimate we could find in the abstract combinatorial setting. In the next subsection we will consider polymers living on finite dimensional lattices; in that situation the number of polymers that can be incompatible with a given one is limited. This will lead to the sharper bounds of Theorems 3.12 and 3.13 which then allow to prove convergence of cluster expansions uniformly in the volume.

b) Application of the formalism to lattice gauge theories : Convergence

We will describe now how the computation of expectation values in lattice gauge theories can be done using the general formalism just developed. We explained the general idea already at the beginning of this chapter. Let us describe the procedure in a more precise way.

First consider again the "high temperature" expansion. A useful thing is to express expectation values by "partition functions", for instance

$$\langle A\rangle_\Lambda = \frac{d}{d\alpha} \log\langle(1+\alpha A)e^{-\mathring{S}}\rangle_{0,\Lambda}\Big|_{\alpha=0}$$

where $\mathring{S} = \mathring{S}_H + S_F + S_{Y.M.}$ if all the fields are present, otherwise we simply omit the corresponding terms.

Similarly we can express truncated expectations, for instance

$$<A;B>_\Lambda \equiv <AB>_\Lambda - <A>_\Lambda <B>_\Lambda$$
$$= \frac{\partial^2}{\partial\alpha\partial\beta} \log<(1+\alpha A+\beta B)e^{-\mathring{S}}>_{0,\Lambda}\Big|_{\alpha=\beta=0} \tag{3.27}$$

So it is sufficient to study "modified partition functions" like

$$Z_\Lambda(A) \equiv <A\, e^{-\mathring{S}}>_{0,\Lambda} \quad . \tag{3.28}$$

Now as explained above $e^{-\mathring{S}}$ has the form $\prod_{b\in\mathcal{B}_\Lambda} (1+\rho_b)$ where $\mathcal{B}_\Lambda$ is the set of "bonds" (links and plaquettes on which fields are coupled).

Expanding the product

$$\prod_{b\in\mathcal{B}_\Lambda} (1+\rho_b) = \sum_{\mathcal{B}\subset\mathcal{B}_\Lambda} \prod_{b\in\mathcal{B}} \rho_b$$

and breaking each product up into "connected" factors corresponding to polymers γ (depending of course on A) we obtain

$$Z_\Lambda(A) = \sum_{\Gamma\subset\Gamma_{A,\Lambda}} \prod_{\gamma\in\Gamma} <A^{n(\gamma)} \prod_{b\in\gamma} \rho_b>_0 \times$$
$$\times \prod_{\substack{i<j \\ \gamma_i,\gamma_j\in\Gamma}} (1+g(\gamma_i,\gamma_j)) \tag{3.29}$$

where

$$n(\gamma) = \begin{cases} 1 & \text{if } \gamma \text{ connects to the support of } A \\ 0 & \text{otherwise} \end{cases}$$

Of course there will be at most one such γ in a set of polymers Γ that is compatible, if we define any two polymers γ ,γ' as incompatible whenever both have any variable in common with A .

In (3.29) we recognize the general form (3.2) if we make the identification

$$z_A(\gamma) \equiv <A^{n(\gamma)} \prod_{b\in\gamma} \rho_b>_0 \quad . \tag{3.30}$$

Using the formalism of the previous section we can write

$$\log Z_\Lambda(A) = \sum_X z_A^X \frac{a(X)}{X!} \tag{3.31}$$

and by (3.26)

$$\langle A\rangle_\Lambda = \sum_X \frac{d}{d\alpha} z_{\alpha A}^X \Big|_{\alpha=0} \frac{a(X)}{X!} \quad , \quad \text{or}$$

$$\langle A\rangle_\Lambda = \Sigma' \; \phi_A^T(X) = \Sigma' \; z_A^X \frac{a(X)}{X!} \tag{3.32}$$

where ' is summation over the terms linear in A . Similarly

$$\langle A_1;A_2;\ldots;A_n\rangle_\Lambda = \Sigma' \; z_{A_1\ldots A_n}^X \frac{a(X)}{X!} \tag{3.33}$$

where the sum Σ' is over the terms linear in $A_1\ldots A_n$. (3.32) and (3.33) are high temperature cluster expansions on standard observables.

There is another kind of observable that is important : The so-called disorder variables, which for the case of gauge theories are known in d = 3 as "monopoles" and in d = 4 as "'t Hooft loops" (both kinds were introduced by 't Hooft [26] ; see Chapter 2.b). One way to introduce them is the duality transformation; in this way they are defined as the duals of order variables ("spin" in d = 3 , Wilson loop in d = 4). In Chapter 2.b we gave a direct definition which we recall :

Select a set of plaquettes *S that is dual to a curve joining two points x,y in d = 3 or to a "sheet" bordered by a closed loop in d = 4 (in d = 2 *S may be any set of plaquettes).

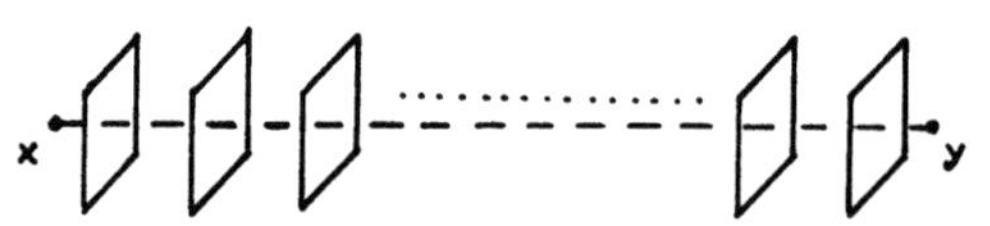

The set *S for d = 3

Modify the building blocks in the Yang-Mills action corresponding to these plaquettes by replacing

$$\chi(g_{\partial P}) \;\text{ by }\; \chi(\omega g_{\partial P}) = \chi(\omega)\chi(g_{\partial P})/\chi(\mathbb{1})$$

$$(\omega \in \text{Center of } G)$$

on these plaquettes. Define the modified partition function $Z_\Lambda(\omega,S)$ accordingly. Then define the "expectation" of the disorder variable $D_\omega(S)$ corresponding to (ω,S) by

$$<D_\omega(S)>_\Lambda \equiv \frac{Z_\Lambda(\omega,S)}{Z_\Lambda} \quad . \tag{3.34}$$

A cluster expansion can be written down immediately for $\log<D_\omega(S)>$ if we notice that the only difference between the expansions for $Z_\Lambda(\omega,S)$ and $Z_\Lambda(\mathbf{1},\emptyset)$ is in the activities of certain polymers : So we write $z_{\omega,S}(\gamma)$ to make this dependence explicit. We obtain (cf. [38]) :

$$\log<D_\omega(S)>_\Lambda = \sum_X (z^X_{\omega,S} - z^X_{\mathbf{1},\emptyset}) \frac{a(X)}{X!} \tag{3.35}$$

Finally, let us make some remarks about the "low temperature" expansion in "defects" (called "vortices" in $d = 3$, "vortex sheets" in $d = 4$). Let us assume that G is discrete; since we only consider compact groups this means in fact that G is finite. We call a defect network a set S of plaquettes that are assigned non-trivial values $g_{\partial P} \neq \mathbf{1}$, together with those values $g_{\partial P}$ (actually it would be better to write $g_{\partial P,x}$ because the starting point x will matter).

This information is equivalent to giving all holonomy operators $g(C_x)$ for all closed loops C_x starting at any point x.

A partition function or modified partition function is then expanded into contributions with fixed defect networks D :

$$Z_\Lambda = \sum_D Z_\Lambda^{(D)} \quad . \tag{3.36}$$

Within each term $Z_\Lambda^{(D)}$ we have to expand the Higgs and Fermion actions as before. The factorization properties dictate the polymer structure : A polymer Λ will consist of a co-connected (connected on the dual lattice) set of defects together with a set of Higgs and Fermions bonds "linked" to it in the following way : Gauge invariance shows that only closed loops of links occur (provided the representations of G that occur have the property (R)); since defects form lines in $d = 3$ and sheets in $d = 4$ the usual linkage relation applies. For the $d = 3$ Higgs model with $G = \mathbb{Z}_2$ all this has been worked out in [41]. The cluster expansions (3.33) and (3.35) can be interpreted as low temperature expansions as well; of course the definitions of polymers and their activities are different now.

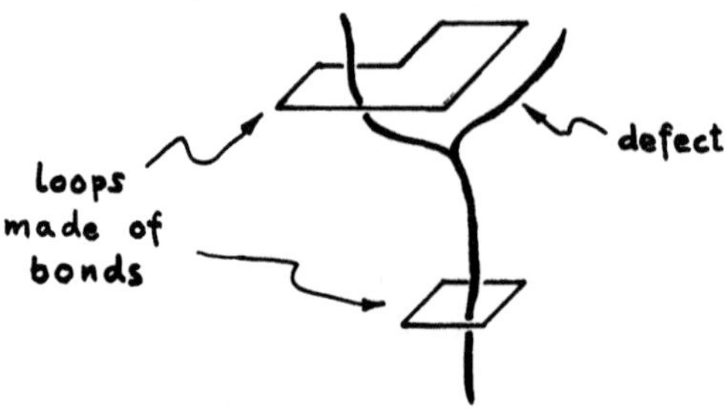

A polymer in the "low temperature region" $(d = 3)$

Finally we want to show that the cluster expansions converge uniformly in the volume, provided the activities are small enough. This could not have been done in the general (dimension-independent) framework of the preceding section, but the dimension enters in an essential way only in one place : The number of polymers that can be mutually incompatible ("overlapping") which occurs in the estimate of $a(X)$ (Cor. 3.9).

Def. 3.10 : By $|\gamma|$ we denote the number of bonds in the polymer γ, considered as a geometric object (i.e. subset of $\mathbb{Z}^d$).

Lemma 3.11 : Let γ be given. The number of polymers of size s that are incompatible with γ is bounded by $|\gamma| C^s$ where C depends on the dimension and the type of polymer considered.

Proof : This is well known; it follows for instance from the solution of the Königsberg bridge problem:

A polymer γ' is a set of bonds, connected in some sense, and picking a bond in γ as a starting point we can move through γ in a path that hits each bond at most twice. The number of such paths of length L is bounded by $C^{L/2} \leq C^s$. The factor $|\gamma|$ comes from the freedom to choose a starting point. □

Theorem 3.12 : (cf. Malyhshev [42]) : There is a constant $K \leq 4/e + \log C$ such that

$$\prod_\gamma C^X(\gamma)^{X(\gamma)} \leq X!\, e^{K \sum_\gamma |\gamma| X(\gamma)}$$

Proof : We set, writing n for $n(X)$:

$$\sum_{|\gamma|=r} X(\gamma) \equiv n_r \tag{3.37}$$

$$\sum_r \frac{n_r}{n} r \equiv d \tag{3.38}$$

(the "mean size" of the polymers in X). Furthermore

$$r_o \equiv \inf_{\gamma} |\gamma| \tag{3.39}$$

The previous lemma says

$$- \sum_{|\gamma|=r} g(\gamma,\gamma') \leq |\gamma'| C^r$$

and therefore

$$\sum_{|\gamma|=r} C^X(\gamma) < -\sum_{\substack{\gamma' \\ |\gamma|=r}} g(\gamma,\gamma')X(\gamma') \leq ndC^r \tag{3.40}$$

Now by the arithmetic-geometric mean inequality

$$\begin{aligned} b(X) &\equiv \sum_{\gamma} X(\gamma)\log C^X(\gamma) - \sum_{\gamma} X(\gamma)\log X(\gamma) \\ &= \sum_{r>r_o} n_r \sum_{|\gamma|=r} \frac{X(\gamma)}{n_r} \log \frac{C^X(\gamma)}{X(\gamma)} \\ &\leq \sum_{r\geq r_o} n_r \log \sum_{|\gamma|=r} \frac{1}{n_r} C^X(\gamma) \qquad (3.41) \\ &\leq \sum_{r\geq r_o} n_r \log \frac{ndC^r}{n_r} \end{aligned}$$

where we used (3.40) in the last step. Rewriting this we get

$$\frac{1}{nd} b(X) \leq \log C + \frac{\log d}{d} - \frac{1}{d} \sum_{r\geq r_o} \frac{n_r}{n} \log \frac{n_r}{n} \tag{3.42}$$

Since

$$\log X! \geq -n + \Sigma X(\gamma)\log X(\gamma) \tag{3.43}$$

what we have to estimate is

$$\tilde{K} \equiv \frac{1}{nd} \sup_{n,d \text{ fixed}} \left(\sum_\gamma X(\gamma) \log C^{X}(\gamma) - \log X! \right)$$

$$\leq \sup_{n,d \text{ fixed}} \left(\frac{1+\log d}{d} + \log C - \frac{1}{d} \sum_{r \geq r_o} \frac{n_r}{n} \log \frac{n_r}{n} \right) \tag{3.44}$$

This reduces to an entropy estimate for an ideal Bose gas. We have to maximize the entropy per particle

$$S \equiv - \sum_{r \geq r_o} \frac{n_r}{n} \log \frac{n_r}{n}$$

under the constraints

$$\sum n_r = n \quad , \quad \frac{1}{n} \sum r n_r = d$$

(d plays the role of internal energy per particle).

It is well known and easy to see that this leads to a Gibbs distribution

$$\frac{n_r}{n} = e^{-\beta(r-r_o)} (1-e^{-\beta})$$

which gives

$$S = - \log(1-e^{-\beta}) + \frac{\beta e^{-\beta}}{1-e^{-\beta}} \tag{3.45}$$

$$d = r_o + \frac{e^{-\beta}}{1-e^{-\beta}} \tag{3.46}$$

This leads very easily to the estimate (using $r_o \geq 1$)

$$S/d \leq \frac{2}{e}$$

[If we want to get better constants we need a little more work :
S is a concave function of d (the specific heat is non-negative)

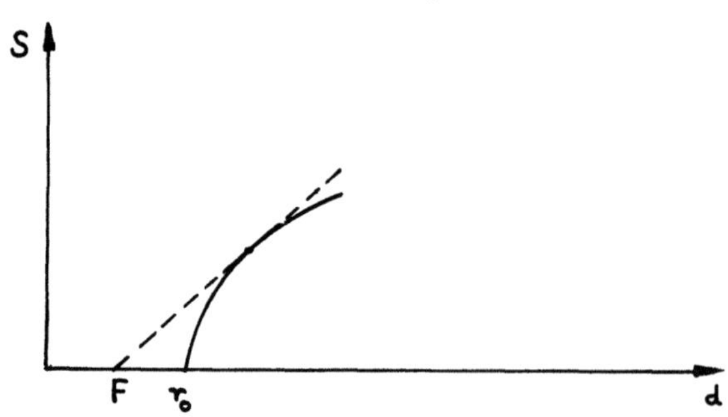

Therefore S/d is maximal when the "free energy" $F = d - \frac{S}{\beta} = r_o + \log(1-e^{-\beta})$ vanishes, i.e. for $\beta = \beta_o = -\log(1-e^{-r_o})$;

$$\sup S/d = \frac{r_o e^{r_o} - (e^{r_o}-1)\log(e^{r_o}-1)}{e^{r_o}+r_o-1} \leq r_o e^{-r_o}$$

$$- (1-e^{-r_o})\log(1-e^{-r_o}) \quad]$$

Looking at (3.44) and using $\frac{1+\log d}{d} \leq \frac{2}{e}$ we see that we can choose (for $r_o \geq 1$)

$$K = \frac{4}{e} + \log C$$ □

Remark : This proof shows that for $r_o \geq 1$ K can be chosen smaller, but never smaller than $\log C$.

Theorem 3.12 implies the following convergence result :

Theorem 3.13 : Let $|z(\gamma)| \leq \exp(-b|\gamma|)$, $b > b_o = 2+K+\log C$, X_o a fixed set of polymers

$$|X_o| = \sum_{\gamma \in X_o} |\gamma| \quad .$$

Then there is a constant c such that

$$\sum_X{}' z^X \frac{a(X)}{X!} \leq c|X_o|e^{-b}$$

where Σ' is the sum over all X such that $\sum_{\gamma_o \in X_o} g(\gamma,\gamma_o)X(\gamma) \neq 0$.

Proof : By Theorem 3.12

$$\sum_X{}' z^X \frac{a(X)}{X!} \leq \sum_X{}' \prod_\gamma e^{(K-b)|\gamma|X(\gamma)}$$

We rewrite this as

$$\sum_{\ell=1}^{\infty} \sum_{\gamma_1,\dots,\gamma_\ell}{}' \frac{1}{\ell!} \sum_{n_1,\dots,n_\ell \geq 1} \prod_{i=1}^{\ell} e^{-(b-K)n_i|\gamma_i|}$$

$$= \sum_{\ell \geq 1} \sum_{\gamma_1,\dots,\gamma_\ell}{}' \frac{1}{\ell!} \frac{e^{-(b-K)|\gamma_i|}}{1-e^{-(b-K)|\gamma_i|}}$$

$$\leq \sum_{\ell \geq 1} \frac{1}{\ell!} (1-e^{K-b})^{-\ell} \sum'_{\gamma_1,\ldots,\gamma_\ell} e^{-(b-K)|\gamma_i|}$$

where $\sum'_{\gamma_1,\ldots,\gamma_\ell}$ is the sum over all ordered sequences $\gamma_1,\ldots,\gamma_\ell$ such that

$$\sum_{\gamma_o \in X_o} \sum_{i=1}^{\ell} g(\gamma_i,\gamma_o) \neq 0 \quad .$$

(and of course $\gamma_1,\ldots,\gamma_\ell$ correspond to a connected graph).

If we prescribe the sizes $k_1 = |\gamma_1|$, $\ldots, k_\ell = |\gamma_\ell|$ there are by Lemma 3.11 at most

$$|X_o| \; \ell (\sum_{i=1}^{\ell} k_i)^\ell \; C^{\sum_{i=1}^{\ell} k_i} \leq \ell! \; \ell |X_o| \; e^{(1+\log C) \sum_{i=1}^{\ell} k_i}$$

such sequences. So we obtain

$$|\Sigma' z^X \frac{a(X)}{X!}| \leq |X_o| \sum_{\ell \geq 1} \ell \sum_{k_1,\ldots,k_\ell \geq 1} (1-e^{K-b})^{-\ell} e^{(1+K+\log C-b) \sum_{i=1}^{\ell} k_i}$$

$$= |X_o| \sum_{\ell \geq 1} (1-e^{K-b})^{-\ell} \ell (\frac{e^{1+K+\log C-b}}{1-e^{1+K+\log C-b}})^\ell$$

$$\leq |X_o| \sum_{\ell \geq 1} \ell (\frac{e^{-1}}{1-e^{-1}} \frac{1}{1-C^{-1}e^{-2}})^\ell e^{\ell(b_o-b)} \leq$$

$$\leq \text{const.} \; e^{-b+b_o}$$

□

Theorem 3.13 suffices to give convergence of the cluster expansion (3.35). For (3.33) some small additional argument is needed. Since there one of the polymers - call it γ_A - will consist of a number of ordinary polymers together with the support of A , we need an estimate of the number of such configurations, given $|\gamma_A|$. This is easily seen to be bounded by $|A| \; C^{|\gamma_A|}$. Then we may proceed as before. We note the essential consequence of this :

<u>Cor. 3.13'</u> : Let $|z(\gamma)| \leq e^{-b|\gamma|}$, b large enough $(b > 4/e+2+2 \log C)$. Then the cluster expansions (3.33) and (3.35) converge absolutely and uniformly in Λ The limit $\Lambda \nearrow \mathbb{R}^d$ exists and is independent of boundary conditions and the result is analytic in the basic coupling constants in the domain that makes the assumed bound on $z(\gamma)$ true.

<u>Proof</u> : Standard and fairly trivial at this stage. □

Note added in proof.

After completion of the revised version of this manuscript a University of Rome preprint by C. Cammarota appeared ("Decay of Correlations for Infinite Range Interactions in Unbounded Spin Systems", to appear in Comm. Math. Phys.), which follows a strategy similar to the one used here for proving convergence of cluster expansions; his handling of the combinatorics is in some way more efficient. Therefore I want to outline his method, as applied to our situation.

Instead of using our Theorems 3.12 and 3.13 one may proceed as follows : First sum over all multi-indices (clusters) X giving rise to a fixed (connected) graph and then sum over the graphs. This sum over connected graphes may be replaced by first summing over all graphs containing a given tree T and then summing over trees, provided we correct for overcounting by dividing by $|\mathcal{T}(G(X)|$ (cf. Thm. 3.6).

So we obtain

$$\sum_X{}' \frac{a(X)}{X!} z^X = \sum_T \sum_{X:G(X)\supset T} \frac{1}{X!} \frac{a(G(X))}{|\mathcal{T}(G(X))|} z^X$$

and by Theorem 3.6 therefore

$$\Big|\sum_X{}' \frac{a(X)}{X!} z^X\Big| \le \sum_T \sum_{X:G(X)\supset T}{}' \frac{1}{X!}|z^X| = \sum_{n\geq 1} \frac{1}{n!} \sum_{T_n} \omega(T_n)$$

where $\sum_{T_n}$ is over all trees on n vertices and

$$\omega(T_n) \equiv \sum_{X:G(X)\supset T_n} \frac{n!}{X!}|z^X|$$

It is convenient to rewrite this as

$$\omega(T_n) = \sum_{\substack{\gamma_1,\ldots,\gamma_n \\ G(\gamma_1,\ldots,\gamma_n)\supset T_n}} \Big|\prod_{i=1}^n z(\gamma_i)\Big|$$

Recall that Σ' means that one of the γ's has to be incompatible with the fixed set of polymers X_o ; we may therefore estimate $\omega(T_n)$ by assuming that it is γ_1 and multiplying by n . Recall that by Lemma 3.11

$$\frac{1}{|\gamma_o|} \sum_{g(\gamma,\gamma_o)=-1} |z(\gamma)| \le (Ce^{-h})^k = \varepsilon^k$$

Inserting this we obtain

$$\omega(T_n) \leq n \sum_{k_1,\dots,k_n \geq 1} |X_o| \prod_{i=1}^{n} \varepsilon^{k_i} \prod_{i=1}^{n} k_i^{C^{T_n}(\gamma_i)-1}$$

Now we use the following bound that holds for $\varepsilon < e^{-1}$:

$$\sum_{k \geq 1} \varepsilon^k k^p \leq p! \frac{\varepsilon}{1-\varepsilon}$$

to obtain

$$\omega(T_n) \leq \prod_{i=1}^{n} (C^{T_n}(\gamma_i)-1)! \left(\frac{\varepsilon}{1-\varepsilon}\right)^n |X_o| n$$

By Cayley's formula the number of trees on $\{\gamma_1,\dots,\gamma_n\}$ with fixed $C^{T_n}(\gamma_i) \equiv d_i$, $i = 1,\dots,n$ is

$$\frac{(n-2)!}{(d_1-1)!\dots(d_n-1)!}$$

and by another simple combinatoric estimate the number of sequences $d_1,\dots,d_n$ with $\sum_{i=1}^{n} d_i = 2n-2$ (as has to be the case here) is bounded by

$$2^{2n-3}$$

Putting everything together we obtain

$$\sum_{n \geq 1} \frac{1}{n!} \sum_{T_n} \omega(T_n) \leq \sum_{n \geq 1} \frac{n(n-2)!}{n!} 4^{n-1} \left(\frac{\varepsilon}{1-\varepsilon}\right)^n |X_o| =$$

$$= - \frac{\varepsilon}{1-\varepsilon} |X_o| \log\left(1 - \frac{4\varepsilon}{1-\varepsilon}\right)$$

and hence convergence of the cluster expansion, provided $\varepsilon < \frac{1}{5}$, or

$$b > \log C + \log 5$$

This is a considerable improvement over the bound given in Cor. 3.13' and again it may be improved even more by taking advantage of a minimal polymer size $r_o > 1$.

c) Results : Consequences of Cluster Expansion Convergence

The convergence of cluster expansions allows to deduce upper and lower bounds on expectations of various types of observables such as Wilson loops, 't Hooft loops and others from bounds on polymer activities. We find various regions in which different types of expansions converge and the qualitative behavior of the expectation values is often different. This suggests phase diagrams for the various models which we give at the end of this chapter. We first focus on pure Yang-Mills theories.

Theorem 3.14 : In a pure lattice Yang-Mills theory there is a "high temperature" phase (g_o^2 large enough) which has the following properties :

(1) Tree graph decay (cf. [44])

$$|\langle A_1;\ldots;A_n\rangle| \leq C(\|A_1\|,\ldots,\|A_n\|)\exp(-mt(A_1,\ldots,A_n))$$

where $C(.,\ldots,.)$ depends on some translation invariant norms of $A_1,\ldots,A_n$. $t(A_1,\ldots,A_n)$ is the length of the smallest tree connecting $A_1,\ldots,A_n$.

(2) The Wilson loop follows the area law if it belongs to a representation of G that does not represent the center of G trivially. More explicitly let C be a closed loop, $[g_C]$ the conjugacy class of the corresponding holonomy operator, χ_τ a character with nontrivial dependence on the center of G, $A(C)$ the minimal number of plaquettes of a surface having C as its boundary.

Then

$$|\langle\chi_\tau(g_C)\rangle| \equiv |\langle W_\tau(C)\rangle| \leq \text{const } e^{-\alpha_\tau A(C)} \qquad (\alpha_\tau > 0) .$$

(3) The 't Hooft loop in $d = 4$ follows the perimeter law, i.e.

$$|\langle D_\omega(C)\rangle| \geq \text{const} \exp(-\beta_\omega|C|) ;$$

$$\text{in } d = 3 , \ |\langle D_\omega(S_{xy})\rangle| \geq \text{const.} > 0 .$$

(4) If the representation τ appearing in the Wilson loop and the representation σ appearing in the action are such that for some $n \in \mathbb{N}$ $\tau \times \sigma^n \times \bar{\sigma}^n$ contains the trivial representation, then the Wilson loop follows the perimeter law

$$|\langle W_\tau(C)\rangle| \geq \text{const } e^{-\beta_\tau|C|} \qquad (d \geq 3)$$

(Example : $G = SU(3)$, σ the fundamental representation, τ the octet representation).

Remarks :

(2) was noted already by Wegner [5] for $G = \mathbb{Z}_2$ and stated more generally (without the qualification referring to the center) by Wilson [6]; a proof was given in [19].

(3) was stated by 't Hooft [26]; for $G = \mathbb{Z}_2$ already by Wegner [5].

(4) was stated by Glimm and Jaffe [45].

(1) expresses of course the existence of a mass gap.

(2) expresses "confinement of quarks" as explained earlier.

(3) expresses nonconfinement of magnetic monopoles in $d = 4$ and some kind of "condensation of defects" in $d = 3$.

(4) expresses the fact that for instance external gluon sources are screened by the dynamical gluons; it is believed that this color shielding mechanism works to prevent physical gluons from being seen (cf. Mack [86]).

The leading behavior of m (the "mass"), α (the "string tension") and β can be easily computed : m , α are $\mathcal{O}(\log \operatorname{Re} g_o^{-2})$, β is $\mathcal{O}(\operatorname{Re} g_o^{-2})$.

Münster [46] has calculated α up to 7th order in $\frac{1}{g_o^2}$.

Proof : We give the proof for the Wilson action; the generalization to other actions is straightforward.

(1) follows from the remark made after eq.(3.33) expressing the fact that in the cluster expansion (3.33) only multi-incides X contribute that have $X(\gamma) > 0$ at least for a "tree" of polymers connecting the supports of $A_1,\dots,A_n$ and the observation that

$$|z^X_{A_1,\dots A_n}| \leq \exp\{-b_{A_1,\dots,A_n} \sum_\gamma X(\gamma)|\gamma|\} ,$$

with $b_{A_1,\dots,A_n} = \mathcal{O}(\frac{1}{g_o^2})$, together with Theorem 3.13.

(2) follows by the same method from

Lemma 3.15 : (1) $z_{W_\tau(C)}(\gamma) = 0$

unless γ contains a surface having C as its boundary.

(2) $|z_{W_\tau(C)}(\gamma)| \leq \exp(-bA(C))$.

Proof : (1) follows from

$$\int \prod_{P\in\gamma} \rho_{\partial P}\, W_\tau(C) \prod_{\langle xy\rangle} dg_{xy} = 0 \tag{3.48}$$

unless γ is as specified. To prove (3.48) we use the following simple formula (cf. [47])

$$\int_G dg\, F(g) = \int_G dg \int_Z d\omega F(\omega g)$$

where G is any group and Z its center. So (3.48) follows from the stronger statement

$$\int \prod_{P\in\gamma} \rho_P(\omega_{\partial P} g_{\partial P}) \chi_\tau(\omega_C) \prod_{\langle xy\rangle} d\omega_{xy} = 0 \tag{3.49}$$

unless γ is as specified.

We insert the Fourier expansion of $\prod_{P\in\gamma} \rho_P(\omega_P g_P) \equiv R(\omega)$ considered as a function on the center of $G^{|\gamma|}$. We obtain a linear combination of terms of the form

$$\int \prod_{P\in\gamma} \chi_{\tau_P}(\omega_{\partial P}) \prod_{b\in C} \chi_\tau(\omega_b) \prod_{\langle xy\rangle} d\omega_{xy} \tag{3.50}$$

which vanish unless each link contributes the trivial representation.

If we define the "boundary" $\partial\gamma$ to be the set of links where

$$\prod_{\substack{P \\ \langle xy\rangle\in P}} \chi_{\tau_P}(\omega) \neq 1 ,$$

we see that

$$\prod_{P\in\gamma} \chi_{\tau_P}(\omega_{\partial P}) = \prod_{\langle xy\rangle\in\partial\gamma} \Big(\prod_{\substack{P \\ \langle xy\rangle\in P}} \chi_{\tau_P}(\omega_{\langle xy\rangle})\Big)$$

("Stokes' theorem").

We see that (3.50) can only be different from 0 if $\partial\gamma = C$, from which (1) follows.

(2) follows again from $|z_{W(C)}(\gamma)| \leq \exp(-b|\gamma|)$ and (1).

End of proof of Lemma 3.15. □

Remark : This shows also that $\alpha_\tau = \infty$ if τ represents the center nontrivially but the representation σ occurring in the action represents it trivially.

We continue the proof of Thm. 3.14 :

To prove (3) we have to use the expansion (3.35). consider $d = 3$, $S = {}^*S_{<xy>}$ a set of plaquettes as indicated in the picture (x,y are sites in the dual lattice).

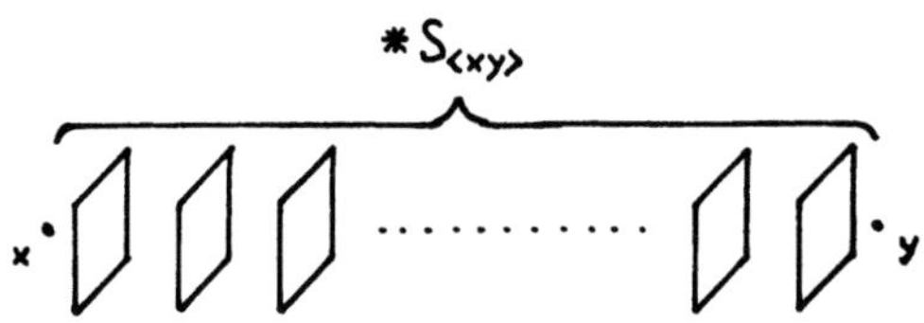

We write in self-explanatory notation

$$<xy> = \partial S_{<xy>}$$

(i.e. the pair $<xy>$ is the boundary of the string $S_{<xy>}$). Notice that the change $g_{\partial P} \to \omega g_{\partial P}$ for $P \in {}^*S_{<xy>}$ can be produced by multiplying some link variables g_{xy} by ω provided γ does not intersect the coboundary of ${}^*S_{<xy>}$ (consisting of the elementary cubes centered at x and y). So by the invariance of the Haar measure for such a polymer γ

$$z_{\omega,S}(\gamma) = z_{\mathbf{1},\emptyset}(\gamma) .$$

Therefore the leading contributions to (3.35) will come from multi-indices ("clusters") $\delta_{\gamma\gamma_x}$ or $\delta_{\gamma\gamma_y}$ that vanish everywhere except on a single polymer γ_x or γ_y consisting of the elementary cube dual to x or y, respectively; $\delta_{\gamma\gamma_x}(\gamma_x) = 1$ and $\delta_{\gamma\gamma_y}(\gamma_y) = 1$.

So the leading terms have the asserted behavior. To get a lower bound, we need a little result about the remainder.

Prop. 3.16 : Let $\mathrm{Re}\,\frac{1}{g_o^2}$ be large enough. Then

$$R \equiv \Big| \sum_{X \neq \delta_{\gamma\gamma_x}, \delta_{\gamma\gamma_y}} (z^X_{\omega,S} - z^X_{\mathbf{1},\emptyset}) \frac{a(X)}{X!} \Big|$$

$$\leq |z_{\omega,S}(\delta_{\gamma\gamma_x}) - z_{\mathbf{1},\emptyset}(\delta_{\gamma\gamma_x})| c e^{-b}$$

$(b = \mathcal{O}(\mathrm{Re}\log g_o^{-2}))$.

Proof : This is a simple variation of the proof of Theorem (3.13). We only have to note one thing : Let $\omega_{\partial P} = \omega$ for $P \in {*S}$ and $\omega = \mathbf{1}$ for $P \notin {*S}$. Then

$$z_{\omega,S}(\gamma) - z_{\mathbf{1},\emptyset}(\gamma) = \int dg \Big[\prod_{P\in\gamma} \big(e^{\frac{1}{2g_o^2\chi(\mathbf{1})}\chi(g_{\partial P})\chi(\omega_{\partial P})} - 1\big) - \prod_{P\in\gamma} \big(e^{\frac{1}{2g_o^2}\chi(g_{\partial P})} - 1\big)\Big] \quad . \qquad (3.51)$$

We now telescope the difference of products and use the fact that for g_o^2 large there is a $c > 1$ such that

$$\Big|e^{\frac{1}{2g_o^2}\chi(g)\cdot\frac{\chi(\omega)}{\chi(\mathbf{1})}} - 1\Big| \leq c\,\Big|e^{\frac{1}{2g_o^2}\chi(g)} - 1\Big|$$

to obtain

$$|z_{\omega,S}(\gamma) - z_{\mathbf{1},S}(\gamma)| \leq$$

$$\leq \sum_{P\in *S\cap\gamma} c^{|\gamma|} \int dg \prod_{\substack{P'\in\gamma \\ P'\neq P}} \Big|e^{\frac{1}{2g_o^2}\chi(g_{\partial P'})} - 1\Big| \times$$

$$\times \Big|e^{\frac{1}{2g_o^2}\chi(g_{\partial P})\chi(\omega_{\partial P})/\chi(\mathbf{1})} - e^{\frac{1}{2g_o^2}\chi(g_{\partial P})}\Big| \leq$$

$$\leq \mathrm{const}|\gamma|\Big|\frac{c}{g_o^2}\Big|^{|\gamma|}\Big(1 - \frac{\chi(\omega)}{\chi(\mathbf{1})}\Big)$$

$$\leq \mathrm{const}'\ \Big|\frac{c}{2}\, g_o^{-2}\Big|^{|\gamma|}\Big(1 - \frac{\chi(\omega)}{\chi(\mathbf{1})}\Big) \ .$$

Since $$|z_{\omega,S}(\delta_{\gamma\gamma_x}) - z_{1,\emptyset}(\delta_{\gamma\gamma_x})| = O(g_o^{-6}) \quad ,$$

we conclude

$$|z_{\omega,S}(\gamma) - z_{1,\emptyset}(\gamma)| \leq$$

$$\leq \text{const } |\frac{g_o}{2}|^{-2|\gamma|+6} |z_{\omega,S}(\delta_{\gamma\gamma_x}) - z_{1,\emptyset}(\delta_{\gamma\gamma_x})| c^{|\gamma|} \tag{3.52}$$

To estimate $$z^X_{\omega,S} - z^X_{1,\emptyset}$$

we have to telescope once more : the result is

$$|z^X_{\omega,S} - z^X_{1,\emptyset}| \leq \text{const } |\frac{g_o^2}{2c}|^{-\sum_\gamma X(\gamma)|\gamma|} \times$$

$$\times \; |z_{\omega,S}(\delta_{\gamma\gamma x}) - z_{1,\emptyset}(\delta_{\gamma\gamma_x})| \quad . \tag{3.53}$$

This can be fed into Theorem 3.13 and produces the assertion of Prop. 3.16.

□

Prop. 3.16 ensures that the leading terms are really leading uniformly in the size of S , so the three-dimensional version of Theorem 3.14, (3) is proven.

For $d = 4$ the situation is quite analogous, only the geometry is a little more complicated. The leading terms come now from polymers that are elementary cubes dual to links in C and there are $O(|C|)$ of them. The remainder can be estimated as before.

To prove (4) we use the same technique. Under the assumption made there, there will be a leading polymer γ consisting of a torus containing C , so $|\gamma| = O(|C|)$

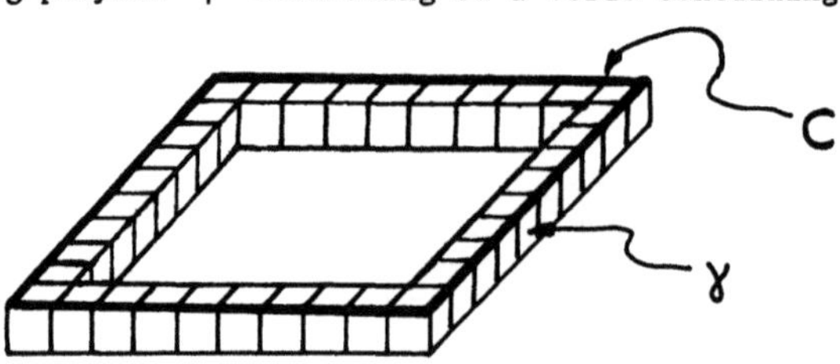

and the leading contribution will be of the form $\exp(-\beta_\tau|C|)$. To estimate the remainder, one has to show that for $\gamma' \supset \gamma$,

$$|z_{W_\tau}(\gamma')| \leq |z_{W_\tau}(\gamma)| (\frac{1}{2g_o^2})^{|\gamma|-|\gamma'|} \tag{3.54}$$

This is clearly true if we arrange for $\rho_{\partial P} \geq 0$ by subtracting a suitable constant from the action, because then

$$z_{W_\tau}(\gamma') = \int \prod_{P\in\gamma} \rho_{\partial P} \prod_{P\in\gamma'\setminus\gamma} \rho_{\partial P}\, dg \leq \Big\| \prod_{P\in\gamma'\setminus\gamma} \rho_{\partial P} \Big\|_\infty z_{W_\tau}(\gamma)$$

(note that g_o^2 has to be taken real here).

This concludes the proof of Theorem 3.14. □

We now turn to the "low temperature" region of pure Yang-Mills theories.

Theorem 3.17 : In a pure lattice Yang-Mills theory with a finite gauge group G there exists a "low temperature" phase (Re g_o^{-2} large enough) with the properties :

(1) Tree graph decay

$$|<A_1;\dots;A_n>| \leq C(\|A_1\|,\dots,\|A_n\|)\exp(-mt(A_1,\dots,A_n))$$

(see Thm. 3.14 for notation).

(2) The Wilson loop follows the perimeter law, i.e.

$$|<\chi_\tau(g_c)>| \equiv |<W_\tau(C)>| \geq \text{const } e^{-\beta_\tau |C|} \quad , \text{ in } d \geq 3 \ .$$

(3) The 't Hooft loop in $d = 4$ has area law decay, i.e.

$$|<D_\omega(C)>| \leq \text{const } e^{-\alpha_\omega A(C)} \quad , \quad (\alpha_\omega > 0)$$

where $A(C)$ denotes the number of plaquettes in the dual lattice in the minimal surface bordered by C . In $d = 3$ point defects cluster exponentially (vortices are massive).

Remarks : (2) was noted for $G = \mathbb{Z}_2$ by Wegner [5] and also discussed as a possibility by Wilson [6]; a proof for $d = 3$ and $G = \mathbb{Z}_2$ was sketched by Gallavotti et al [38]. (3) was stated by 't Hooft [26]. For abelian G it has been known for a while that Theorem 3.17 follows by a duality transformation from Theorem 3.14 (cf. [5]). This is hard to carry out, however, for a nonabelian group (possibly impossible). The interpretation of (2) and (3) is : quarks are not confined, but monopoles are.

Finiteness of G is not really essential, only discreteness; but for infinite discrete G the expansions have to be reorganized somewhat (cf. Section 4c).

Proof : We only give some general ideas since the details are similar as in the proof of Theorem 3.14.

(1) is essentially identical

(2) The leading polymers in $d = 3$ are vortex lines winding around the loop (cf. Mack [48], Göpfert [96]):

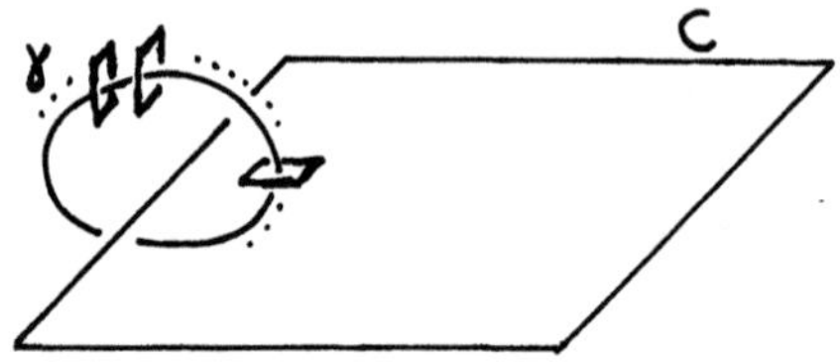

If we use the type of cluster expansion exemplified in (3.35) for the 't Hooft loop, that is, we expand

$$\log Z_\Lambda(W_\tau)-\log Z_\Lambda = \sum_X (z^X_{W_\tau} - z^X_1) \frac{a(X)}{X!} \tag{3.55}$$

There are $\mathcal{O}(|C|)$ leading terms; the estimation of the remainder proceeds as in the proof of part (3) of the previous theorem (actually it is even easier here since no integrations are involved).

(3) We have to adapt the expansion of the type (3.32) for the disorder parameter :

$$\langle D_\omega(S)\rangle_\Lambda = Z_\Lambda(\omega,S)/Z_\Lambda = \frac{d}{d\alpha} \log(Z_\Lambda+\alpha Z_\Lambda(\omega,S))\Big|_{\alpha=o} \tag{3.56}$$

$Z_\Lambda(\omega,S)$ has an expansion in terms of defects similar to $Z_\Lambda(\mathbf{1},S) = Z_\Lambda$ but the constraints are different at the boundary ∂S due to the singular nature of the "gauge transformation" ω . In particular there will always be a defect γ having the same boundary as S .

Expanding the logarithm in (3.56) and differentiating we obtain

$$\langle D_\omega(S)\rangle_\Lambda = \sum_X{}' z^X_{\omega,S} \frac{a(X)}{X!} \tag{3.57}$$

where the sum Σ' is over multi-indices X having the following property :

There is a distinguished polymer (defect) γ_o having $X(\gamma_o) = 1$ that is obtained from an ordinary allowed defect by the singular gauge transformation ω ; all other defects γ with $X(\gamma) > 0$ have to obey the unmodified constraint (i.e. they have to be obtainable from a gauge field).

The distinguished defect produces a factor

$$|z_{\omega,S}(\gamma_o)| \leq e^{-\frac{\beta}{g_o^2}|\gamma_o|} ,$$

where

in $d = 4$, $|\gamma_o| \geq A(C)$; $(C = \partial S)$

in $d = 3$, $|\gamma_o| \geq \text{dist}(x,y)$; $(\langle xy \rangle = \partial S)$

so we obtain an area law in $d = 4$ and exponential clustering in $d = 3$ as asserted. □

Now we want to include matter fields. First we consider Higgs models (no fermions) with so-called complete breakdown of symmetry. This means that the stability group of a point ϕ_o that minimizes $V(|\phi|)$ is trivial. We also make the simplifying (not essential) assumption that $U_H(G)$ acts transitively on that space of minima.

Examples for this situation are

(1) $G = U(1)$, $\phi(x) \in \mathbb{C}$, ϕ transforming under the fundamental representation.

(2) $G = SU(2)$, ϕ in the fundamental representation.

(3) $G = SO(3)$, ϕ consists of a pair of fields in the vector representation; the potential assumes its minimum for an orthogonal pair.

For simplicity let us formally take a limit of V that means restricting ourselves to fields ϕ of fixed length (say $|\phi| = 1$) . This is not essential but reduces technical complications (see [19] for a more general discussion). Then we have the following result.

Theorem 3.18 : In a Higgs model with complete symmetry breakdown and $|\phi| = 1$, the "high temperature" cluster expansion converges in a region $\{\lambda, g_o^2 | \lambda$ and g_o^{-2} small enough or λg_o^2 large enough$\}$.

In this region we have

(1) Exponential clustering.

(2) Area law decay of the 't Hooft loop in $d = 4$, exponential clustering of defects (monopoles) in $d = 3$ for $\lambda > 0$.

(3) Perimeter decay of the Wilson loop for $\lambda > 0$.

Remarks : (1) The most remarkable feature is the extension of the region of convergence of the cluster expansion to small g_o^2 , provided λ is large. This is a sign of the Higgs mechanism. It was proven in [19] and discussed afterwards in [49].

(2) Due to the presence of the Higgs field the disorder variables depend on the whole set S on which the plaquette variables are modified, not only on its coboundary. So the "'t Hooft loop" depends actually on a sheet bordered by the loop.

Proof : The "complete breakdown of symmetry" plus our transitivity assumption mean that the space where the Higgs fields live becomes homeomorphic to G and we may replace the field ϕ by a field taking values in G.

(Fix a $\phi_o \in \mathcal{U}_H$; this determines uniquely a $h_x \in G$ such that $U_H(h_x)\phi_o = \phi(x)$).

Now the Higgs field can be eliminated completely by going to the so-called U-gauge : This means that we make the gauge transformation

$$g_{xy} \to h_x^{-1} g_{xy} h_y$$

$$\phi(x) \to U_H(h_x^{-1})\phi(x) = \phi_o \quad .$$

In terms of the transformed variables the Higgs action is (up to a constant)

$$S_H(\{g_{xy}\}) = \lambda \sum_{<xy>} (1-(\phi_o, U_H(g_{xy})\phi_o)) \tag{3.58}$$

Now $1-(\phi_o, U_H(g_{xy})\phi_o) \geq 0$ and $= 0$ only for $g = \mathbf{1}$ which means that the Higgs action has the effect of increasing the weight of the neighborhood of $\mathbf{1}$ in the space of gauge fields similar to a magnetic field in a spin system. This serves to effectively increase g_o^2, hence the mass generation even for small g_o^2.

Note that S_H does not couple different links; therefore we can include e^{-S_H} in the decoupled expectation and perform the cluster expansion only with respect to the Yang-Mills action. Define a probability measure

$$d\sigma(g) = \frac{1}{C(\lambda)} e^{\lambda((\phi_o, U_H(g)\phi_o)-1)} dg \tag{3.59}$$

and

$$<A>_{o,\Lambda} = \int A \prod_{<xy>} d\sigma(g_{xy}) \quad . \tag{3.60}$$

$$<A>_\Lambda = \frac{1}{Z_\Lambda} <A\, e^{-S_{Y.M.}}>_{o,\Lambda}$$

may then be expanded as before (see eq. (3.32)). The polymers are now just connected sets of plaquettes and the activities are

$$z_A(\gamma) = \langle A \prod_{P\in\gamma} \rho_{\partial P}\rangle_{o,\Lambda} \quad . \tag{3.61}$$

The convergence condition is according to Theorem 3.13

$$|\langle A \prod_{P\in\gamma} \rho_{\partial P}\rangle_{o,\Lambda}| \leq e^{-b|\gamma|} \tag{3.62}$$

with b sufficiently large. Now by Hölder's inequality

$$\begin{aligned} &|\int A \prod_{P\in\gamma} \rho_{\partial P} \prod_{\langle xy\rangle} d\sigma(g_{xy})| \leq \\ &\leq ||A||_\infty (\int |\rho_{\partial P}|^r \prod_{\langle xy\rangle\in P} d\sigma(g_{xy}))^{\frac{|\gamma|}{r}} \end{aligned} \tag{3.63}$$

Here r is a number counting how many different plaquettes can share a link and it is crucial that this number is fixed and does not increase with $|\gamma|$.

Now it is not hard to see that

$$\int |\rho_{\partial P}|^r \prod_{\langle xy\rangle\in P} d\sigma(g_{xy})$$

is $\quad \mathcal{O}(1/g_o^{2r})$ for g_o^2 large, λ small

and $\quad \mathcal{O}((g_o^2\lambda)^{-r}$ for λ large.

which proves the convergence condition stated in the theorem.

For the disorder variables we proceed as in the proof of Theorem 3.14, but now the activities $z_{\omega,S}(\gamma)$ are defined as

$$\begin{aligned} z_{\omega,S}(\gamma) \equiv \prod_{\langle xy\rangle} \int d\sigma(g_{xy}) \prod_{\substack{P\in\gamma \\ P\notin *S}} (e^{\frac{1}{2g_o^2}\chi(g_{\partial P})} -1) \times \\ \times \prod_{P\in\gamma\cap *S} (e^{\frac{1}{2g_o^2}\chi(g_{\partial P}\omega)} -1) \quad . \end{aligned}$$

Since $d\sigma(g) \neq d\sigma(\omega g)$ for $\lambda > 0$ it is no longer true that only for polymers γ intersecting the coboundary of $*S$ $z_{\omega,S}(\gamma) \neq z_{\mathbb{1},\emptyset}(\gamma)$. So there will be two kinds of polymers giving leading contributions : First, as in the pure Yang-Mills case, elementary cubes in the coboundary $\delta *S$ of $*S$, second single plaquettes $P \in *S$.

The second kind of polymers certainly give a negative contribution proportional to the area $|S|$ to the cluster expansion of $\log\langle D_\omega(S)\rangle$ because

$$z_{\omega,\{P\}}(\{P\}) < z_{1,\emptyset}(\{P\}) .$$

To see this, use the Fourier expansion

$$e^{\frac{1}{2g_o^2}\chi(g_{\partial P})} - 1 = \sum_\tau a_\tau\left(\frac{1}{g_o^2}\right)\chi_\tau(g_{\partial P})$$

where $a_\tau\left(\frac{1}{g_o^2}\right) \geq 0$ and

$$\frac{1}{C(\lambda)}\, e^{\lambda((\phi_o, U_H(g)\phi_o)-1)} = \sum_\tau \mathrm{Tr}\, A_\tau(\lambda) U_\tau(g)$$

where $A_\tau(\lambda) \geq 0$. So

$$z_{\omega,\{P\}}(\{P\}) = \sum_{\substack{\tau \\ \tau_1,\ldots,\tau_4}} \chi_\tau(\omega) a_\tau\left(\frac{1}{g_o^2}\right) \times$$

$$\times \int \chi_\tau(g_1 g_2 g_3 g_4) \prod_{i=1}^{4} \mathrm{Tr}\, A_{\tau_i}(\lambda) U_{\tau_i}(g_i) dg_i$$

which equals

$$\sum_\tau \chi_\tau(\omega)\tilde{a}_\tau\left(\frac{1}{g_o^2}\right) \mathrm{Tr}\, A_\tau^4$$

with $\tilde{a}_\tau\left(\frac{1}{g_o^2}\right) \geq 0$. This last expression clearly has its maximum at $\omega = 1$.

So the leading terms give an area law; to see that the nonleading terms do not destroy that we have to make a similar argument as in the proof of Theorem 3.14 (3) □

There is also a "low temperature" regime that can be analyzed :

Theorem 3.19 : In a Higgs model with finite gauge group G (not necessarily with complete symmetry breaking) the "low temperature" cluster expansion converges in a region $\{\lambda, g_o^2 | \lambda$ small and g_o^2 small $\}$.

In this region we have

(1) Exponential clustering

(2) Perimeter decay of the Wilson loops

(3) Area law decay for the 't Hooft loop in $d = 4$; exponential clustering of defects in $d = 3$.

Remark : (1) was proven for $d = 3$, $G = \mathbb{Z}_2$ by Marra and Miracle-Solé [41].

Proof : This is almost routine by now. The polymers consist of connected defect networks in the dual lattice embellished by sets of links in the original lattice winding around them as described earlier.

□

Sometimes there is a third region :

Quasi-Theorem 3.20 : In a Higgs model with a gauge group that need not be discrete, in which a discrete subgroup H contained in the center remains unbroken, there is a convergent "low temperature" expansion in the region $\{\lambda, g_o^2 | \lambda g_o^2$ large enough, λ large $\}$.

In this region there is

(1) Exponential clustering

(2) Perimeter decay of the Wilson loops

(3) Area decay of the 't Hooft loop in $d = 4$, exponential clustering of defects in $d = 3$.

Theorem 3.20' : Under the same conditions as in Quasi-Theorem 3.20 there is a convergent "high temperature" expansion in the region

$$\{\lambda, g_o^2 | \lambda \text{ and } g_o^{-2} \text{ small}\}$$

showing for $\lambda > 0$:

(1) Exponential clustering

(2) Perimeter decay for Wilson loops representing the subgroup H trivially, area decay for all others ("confinement of fractional charges")

(3) In $d = 4$ perimeter decay for 't Hooft loops corresponding to $\omega \in H$, for those corresponding to $\omega \notin H$ area decay; analogous behavior of defects in $d = 3$.

Remarks : This means that under this "partial breakdown of symmetry" the region of Theorem 3.18 breaks up into a "confinement region" (Theorem 3.20') and a "Higgs region" (Quasi-Theorem 3.20). This is more or less known to the experts (see [54] for some numerical evidence) but a detailed proof for Quasi-Theorem 3.20 is still missing. We give an outline of a proof below.

Proof of Theorem 3.20' : Essentially a combination of the proofs of Theorem 3.14 and 3.18.

Sketch of a proof of Quasi-Theorem 3.20 : We use the "unitary gauge" again. Here the Higgs field lives in the coset space (factor group) $F = G/H$; a measure $d\sigma(g)$ can still be defined as before, but instead of favoring the vicinity of the identity $\mathbb{1} \in G$, it favors the vicinity of H .

We break up G accordingly into a number of subsets : First we choose a sufficiently small neighborhood U of the identity; U has to be invariant under conjugation and we require $hU \cap h'U = \emptyset$ for any $h,h' \in H$, $h \neq h'$. $E \equiv G \setminus \bigcup_{h\in H} hU$ is called the exceptional set.

To define polymers we first have to partition the field configurations coarsely according to the partition of G . We characterize classes of field configurations by

(a) The set of links $\langle xy \rangle$ for which $g_{xy} \in E$.

(b) The set of "H-defects" characterized by the set of plaquettes for which $g_{\partial P}$ lies in one of the sets hU for $h \neq \mathbb{1}$. We assume that U is so small that a plaquette P cannot belong to an H-defect if all four link variables $g_{\langle xy \rangle}$ ($\langle xy \rangle \in P$) are in the same set hU .

For any plaquette P not belonging to the set of H-defects $\exp(\frac{1}{2g_o^2}\chi(g_P))$ will be likely (with respect to $d\sigma$) to be near 1 , so we make the usual "high temperature" expansion for these plaquettes. The crucial point is to understand the factorization properties that arise.

Let us look at "empty" regions of the lattice, that is regions in which all $g_{\langle xy \rangle}$ are "near" H and the integrand $\exp(\frac{1}{2g_o^2}\chi(g_P))$ is replaced by 1 . Let us denote by h_{xy} the element of H that is near g_{xy} (i.e. $g_{xy} \in h_{xy}U$) . In our empty region the configuration $\{h_{xy}\}$ is determined up to H-gauge transformations. We can fix this H-gauge in an arbitrary way without affecting anything. This shows that contributions from two sets that are separated by an "empty" region factorize.

We should stress that a set in order to qualify as an "empty region" has to be a face-connected set of cubes in $d = 3$ and a face-connected set of hypercubes in $d = 4$.

A polymer is then given by a co-connected set of H-defects (i.e. their duals are connected)together with a set of plaquettes that are not separated by an empty region.

It is clear that the region of parameters is defined in such a way that the polymers acquire a small activity. So the proof of Quasi-Theorem 3.20 should go through without any difficulty. But it would be useful to work out the proof in full detail; we leave it here as an exercise to the reader.

□

<u>Remark</u> : This kind of combined low and high temperature expansion is reminiscent of the mean field expansion of Glimm, Jaffe and Spencer [50], but of course it is much simpler in this lattice system.

In the following chapter we will come back to the Higgs models and discuss in more detail the abelian models, including the notion of θ-states.

We close this discussion of Higgs models with pictures of the expected phase diagrams for finite gauge group G :

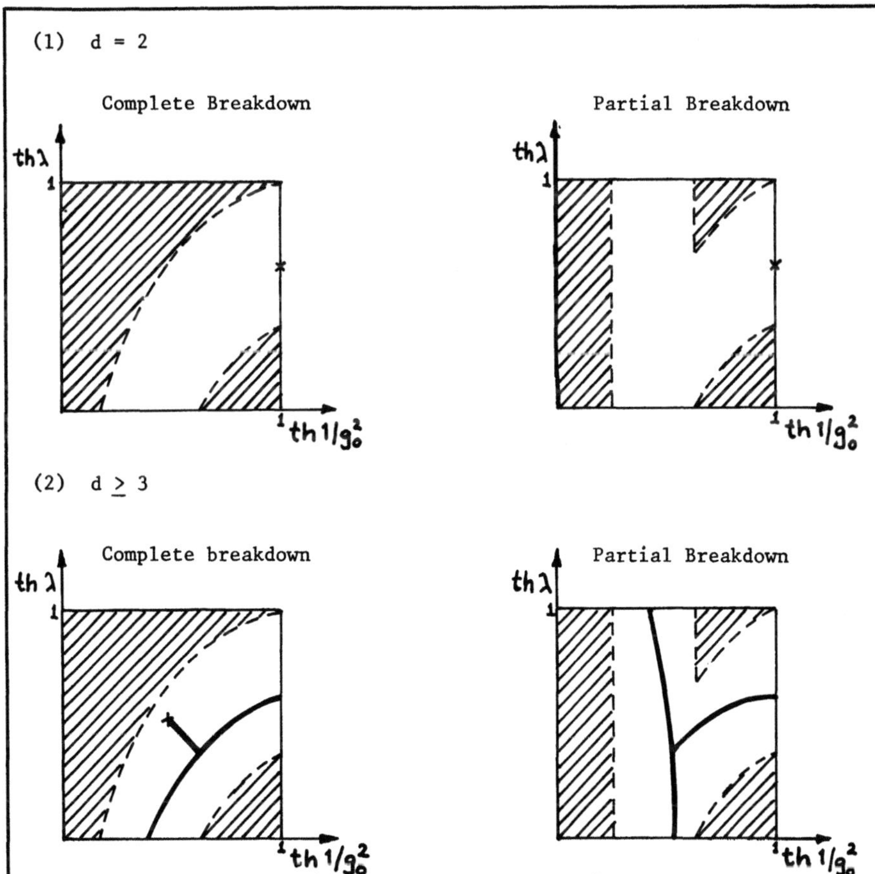

Shaded regions : Convergent expansions

Solid lines : Expected phase transitions

Crosses : Critical points

Remarks : The "spoke" in the left picture ending in a critical point is suggested by numerical results for the $d = 3$, $G = \mathbb{Z}_2$ model [54].

The use of duality can enlarge the shaded regions somewhat in abelian theories $(d = 3)$.

For certain models $(d = 3$, $G = \mathbb{Z}_n$, $n \geq 5)$ the phase boundaries widen to become an intermediary Coulombic phase [61,87].

Last but not least, let us turn to fermion models. From the point of view of physics they are of course the most interesting ones. The results obtained so far by expansion methods leave a larger part of parameter space uncharted, however, and a deeper understanding would certainly be desirable.

Theorem 3.21 : In a Fermion-gauge model (such as lattice QCD) with a faithful representation U_F of G , the "high temperature" cluster expansion converges in a region $\{\kappa, g_o^2 \mid g_o^2$ large and κ small $\}$.

In this region there is

(1) Exponential clustering

(2) Perimeter decay of the Wilson loops

(3) Area decay of the 't Hooft loops in $d = 4$, exponential clustering of defects in $d = 3$.

Remarks :

(1) A similar convergent expansion was discussed by Gawędzki [52] and by Challifour and Weingarten [53] (their criterion for quark confinement does not seem to be appropriate, however).

(2) The perimeter law for the Wilson loop should not be interpreted as breakdown of quark confinement but rather as a sign of "hadronization" : A widely separated pair of external charges will polarize the vacuum to create quark-antiquark pairs that subsequently form hadrons; one pair is used to shield the external charges. This means that the Wilson loop does not give a criterion for confinement when the external charges can be screened by matter fields. Unfortunately we do not have a replacement for this criterion of comparable simplicity.

Note added : A criterion has been proposed recently by Mack and Meyer [94].

(3) The smaller domain of convergence compared to Theorem 3.18 is due to the non-existence of a "unitary gauge" that decouples the expectation at $g_o^2 = \infty$.

Proof : This is straightforward. The polymers will be connected sets of plaquettes and links. For a detailed proof it is useful to use the norms for fermion functions introduced in Section 2. In order to get a good region of convergence, it is probably essential to do the integration over the gauge field first, when one computes the activity of a polymer, because the orthogonality of different representations will produce cancellations in expressions like

$$\int \exp\{\frac{1}{2}\kappa(\bar{\psi}(x)U(g_{xy})\Gamma_{xy}\psi(y)+\bar{\psi}(x)U(g_{xy}^{-1})\Gamma_{yx}\psi(y))\}dg_{xy}$$

It would be worthwhile to check the estimates numerically in order to see whether convergence can be shown for $|\kappa| > \kappa_C$, where κ_C corresponds to mass zero for the free lattice fermion field. This would show a mass generation by the coupling to the gauge field and be a hint of confinement.

□

Theorem 3.22 : In a Fermion-gauge model with discrete gauge group G the "low temperature" cluster expansion converges in a region $\{\kappa, g_o^2 | \kappa$ and g_o^2 small$\}$.

In this region we have

(1) Exponential clustering

(2) Perimeter decay of the Wilson loop

(3) Area decay of the 't Hooft loop (i.e. confinement of external monopoles) in $d = 4$; exponential clustering of defects in $d = 3$.

Proof : Essentially identical to the proof of Theorem 3.19.

□

We summarize the situation again in a picture:

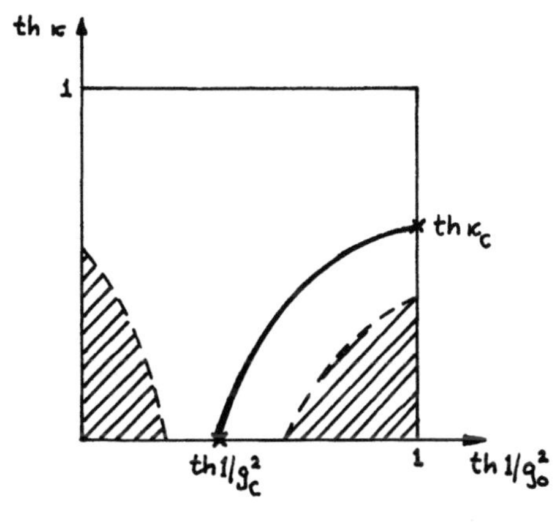

Fermion gauge model with discrete gauge group

The phase boundaries are even more hypothetical than in the Higgs model; the only sure thing is that for $g_o^2 = \infty$, i.e. the free fermion field, there is a critical point at $\kappa = \kappa_C$ $(=\frac{M}{dr}-1$ for $\theta = 0)$ and that there is a critical gauge coupling $g_o = g_c$ for $\kappa = 0$.

Note added : An extensive discussion of the expected phase structure of fermion gauge theories can be found in Kawamoto's work [97].

There are two important things that are insufficiently understood :

(1) Which criterion can replace the Wilson criterion as a signal of confinement ? The first thing coming to mind, namely the 't Hooft loop, does not seem to contain more information : In the presence of matter fields it has the fatal tendency to always follow the area law just as the Wilson loop always tends to show the perimeter law. (See however Mack and Meyer [94] for a proposal of a disorder parameter to distinguish phases).

Of course the "confinement dogma" says : "Check whether the pure Yang-Mills theory shows area decay for the Wilson loop; if so, the full theory will confine quarks". This may be true but it would be nice to know.

(2) Does the theory produce a mass gap for $|\kappa|>|\kappa_c|$ provided g_o^2 is large ? This seems necessary for the confinement of light quarks. Understanding this would also shed more light on the difference between Higgs and Fermion models and thereby on the difference between "screening" (or "bleaching") and confinement (see [86] for a discussion of these concepts).

4. FURTHER DEVELOPMENTS

In this section we describe a selection of results of some physical interest that have been obtained by various people.

The selection is necessarily subjective; in part it is also dictated by the limited space (and time) available. So I cannot discuss in detail the very interesting and sophisticated work by Durhuus and Fröhlich [66]; see however the remarks made in Section 4f ("roughening").

Another important development not treated here is the beautiful work of Göpfert and Mack [93] that has appeared after these lectures were given. They prove that the U(1) model in dimension 3 (with Villain action) shows confinement in the sense of Wilson at all couplings and, maybe more surprisingly, that the ratio of string tension to mass gap ("glueball mass") goes to ∞ when the coupling g_o^2 goes to 0 . The proof uses a remarkable combination of renormalization group ideas with techniques originating in Constructive Quantum Field Theory.

a) Abelian Higgs Models in Two Dimensions : Θ-Vacua, Phase Transition and Confinement of Fractional Charges.

As these headlines indicate , two-dimensional abelian models show some of the fancier features that are believed to be essential in four-dimensional nonabelian theories. One reason for this analogy is of a topological nature : Two-dimensional abelian theories may have a nontrivial topological charge, namely the first Chern class, whereas four-dimensional nonabelian theories may have a non-trivial second Chern class (cf. appendix). The "confinement of fractional charges" in two dimensions is, however, due to a much simpler mechanism (namely an unscreened linear Coulomb potential) than the phenomenon of the same name in four dimensions.

We consider a U(1) Higgs model; because of the abelian nature of the model we can use a Gaussian action for the gauge field (a possibility not discussed in Section 1). For the coupling of Higgs and gauge fields we use a Villain type action. So the total action is $S_o = S_{Y.M.G.} + S_H$

$$S_{Y.M.G} \equiv \frac{1}{2g_o^2} \sum_P \varphi_{\partial P}^2 + \frac{1}{2}\mu^2 \sum_{<x,y>} \varphi_{xy}^2 \tag{4.1}$$

$$S_H \equiv - \sum_{<xy>} \log\{ \sum_{n_{xy}} e^{-\frac{\lambda}{2}(\varphi_{xy}+2\pi n_{xy})^2} \, , \tag{4.2}$$

where we parametrized the elements of the gauge group U(1) by angles φ_{xy} in an obvious way. The mass term in (4.1) is a provisional "infrared cutoff"; it may be safely removed later. Formally $\mu^2 = 0$ corresponds to the Villain action (1.8) for the gauge field since the Gaussian might as well be periodized when it is integrated against periodic functions. To understand the action S_H (4.2) one should note that S_H behaves qualitatively like $\lambda_{<xy>} \cos \varphi_{xy}$ (and has the same formal continuum limit); this last expression would be the standard U(1) Higgs action for a Higgs field of modulus 1 in the unitary gauge (cf. (1.9) and the discussion in the proof of Theorem 3.18). Those wo do not like the Gaussian action may instead work with a conventional lattice action; it is essential for the effects to be discussed, however, that the Higgs action has a shorter period (i.e. higher charge) than the Yang-Mills action ((4.1) has "infinite period").

Definition 4.1 : The Θ-states are defined as thermodynamic limits of the states corresponding to the action

$$S_\Theta \equiv S_o + i\Theta \sum_P \varphi_{\partial P}$$

(where the sum is of course over all plaquettes in a consistently chosen orientation and one should use free boundary conditions, i.e. all variables corresponding to the finite lattice Λ are integrated over).

Remark : The point of this definition is of course that the states actually depend on Θ , in fact in a periodic way as we will see. The Θ-states may be interpreted as having a background electric field. By Stokes' theorem the Θ-states are seen to arise from wrapping an "infinitely large Wilson loop" around the system.

To see the effect of the Θ-term it is best to look at a correlation of point "defects" obtained by shifting the plaquette variables φ_{P_1} and φ_{P_2} by χ and $-\chi$, respectively. We denote the corresponding expectation value by $\langle D_\chi(P_1)\, D_{-\chi}(P_2)\rangle_{\Theta,\Lambda}$.

To evaluate it we use a duality transformation [55,56] . For this purpose we note that up to an irrelevant constant factor

$$\sum_{n\in\mathbb{Z}} e^{-\frac{1}{2\sigma}(\varphi+2\pi n)^2} \sim \sum_{n\in\mathbb{Z}} e^{-\frac{\sigma}{2}n^2} e^{i\varphi n} \tag{4.3}$$

The formula (4.3) can be used to Fourier transform e^{-S_Θ} and to carry out the integrations over the angles φ in $\langle D_\chi(P_1)\, D_{-\chi}(P_2)\rangle_{\Theta,\Lambda}$. By a computation one obtains

$$\langle D_\chi(P_1)\, D_{-\chi}(P_2)\rangle_{\Theta,\Lambda} =$$

$$= \frac{1}{Z_{\Theta,\Lambda}} \sum_{\{n_P\}} e^{i\chi(n_{P_1}-n_{P_2})}\, e^{-\frac{g_o^2}{2}\sum_P (n_P+\frac{\Theta}{2\pi})^2}\, e^{-\frac{1}{2\lambda}\sum_{\langle x,y\rangle} n_{xy}^2} \tag{4.4}$$

where $n_{xy} = n_P - n_{P'}$ with P,P' being the plaquettes sharing the link $\langle xy\rangle$ and the limit $\mu^2 \to 0$ has been taken already. (The boundary conditions will be 0-Dirichlet if we start with free ones in the original model. This simply means that $n_P = 0$ outside of Λ). (4.4) clearly shows the physical content of the model : It is in fact a model of integer spins with nearest neighbor ferromagnetic coupling. Θ enters the single spin distribution in a periodic way, it is analogous to a magnetic field. Furthermore we see that for $\Theta = \pi$ the maximum of the single spin distribution is degenerate ($n_P = 0$ or 1) .

For λ small a standard low temperature expansion shows the coexistence of two phases with exponential clustering, very much like in the low temperature region of the Ising model (cf. [57,58]). Of course the coexistence of two phases may be seen by the standard Peierls argument. Presumably there will be a critical point in λ (depending on g_o^2) at which there is no exponential clustering. This can be interpreted as "breakdown of the Higgs mechanism".

For λ large and $g_o^2 > 0$ there will be a high temperature phase with a convergent cluster expansion (cf. Section 3).

Finally for $g_o^2 = 0$, $\Theta = 0$ the model becomes identical to the Villain form of the plane rotator model. This model shows the famous Kosterlitz-Thouless transition [59] which has been rigorously proven to occur by Fröhlich and Spencer [60].

Now let us turn to the "confinement of fractional charges" for $G = U(1)$:

Theorem 4.2 : Let $\varepsilon(\Theta) \equiv \lim_{\Lambda \nearrow Z^2} \frac{1}{|\Lambda|} \times \log \frac{Z_{\Theta,\Lambda}}{Z_{0,\Lambda}}$. Then

1) $\varepsilon(\Theta)$ is periodic with period 2π

2) $\varepsilon(\Theta) \leq 0$; $\varepsilon(\Theta) < 0$ for $\Theta \neq 0 \bmod 2\pi$

3) $|\langle W_{\Theta'}(C)\rangle_\Theta| \leq e^{(\varepsilon(\Theta+\Theta') - \varepsilon(\Theta))A(C)}$ where $W_{\Theta'}(C) = \prod_{\langle xy\rangle \in C} e^{\frac{i\Theta'}{2\pi}\varphi_{xy}}$ and $A(C)$ is the area enclosed by the loop C .

Remark : A somewhat more restricted form of this theorem is proven in [62]; it appears essentially in this form in [58]. 3) expresses confinement of fractional charges if we set $\Theta = 0$.

Proof : The Θ-states have Osterwalder-Schrader positivity in spite of the complex factor appearing in the expectation value.

Furthermore we can use Osterwalder-Schrader positivity of the uncoupled expectation $\langle\ \rangle_o$ to see existence of the limit $\lim_{\Lambda \nearrow \mathbb{Z}^2} \frac{1}{|\Lambda|} \log Z_{\Theta,\Lambda}$ for rectangles Λ : Let Λ be a rectangle of sides L and T and write $Z_{\Theta,LT} \equiv Z_{\Theta,\Lambda}$. Then by Schwarz's inequality it follows (as in the proof of Lemma 2.4) that

$$Z_{\Theta,LT} \leq Z_{\Theta,LT_1}^{1/2} \; Z_{\Theta,L(2T-T_1)}^{1/2}$$

(L,T,T_1 odd).

This means that $\log Z_{\Theta,LT}$ is a convex function of T (and L) and therefore

$$\frac{1}{(L-1)(T-1)} \log \frac{Z_{\Theta,LT} Z_{\Theta,11}}{Z_{\Theta,1T} Z_{\Theta,L1}}$$

is increasing in L and T. Since $\frac{1}{LT} \log Z_{\Theta,LT}$ is clearly bounded uniformly in L and T it is seen readily that its limit as $L,T \to \infty$ exists.

The existence of $\varepsilon(\Theta)$ is thus proven (the structure of this proof can be traced back to Guerra's work [63]).

$\varepsilon(\Theta) \leq 0$ is trivial. The periodicity follows by a duality transformation as in (4.4). (3) follows by iterating Schwarz's inequality with respect to the O.S. inner product.

What's missing so far is the strict inequality in (2). For this we use the correlation inequalities of 2.d) : First we insert a term $\sum'_{\langle xy \rangle} K \varphi^2_{xy}$ $(K > 0)$ into the action where Σ' is over all space-like links. Sending $K \to +\infty$ decreases the covariance and therefore increases $\langle W_{\Theta'}(C) \rangle_\Theta$ by Theorem (2.10) (1). It also eliminates the gauge fields corresponding to space-like links from the system and decouples different "time layers". We increase $\langle W_{\Theta'}(C) \rangle_\Theta$ even more by sending the coupling constant λ of the Higgs action to $+\infty$. This "freezes" the remaining gauge field variables to $2\pi \times$ integer values. We are left with a "stack" of one-dimensional integer spin models and obtain for a rectangular Wilson-loop of sides $2L'$ and $2T'$ in a box of sides $2L$ and $2T$

$$\langle W_{\Theta'}(C) \rangle_\Theta^{\frac{1}{2T'}} \leq \frac{1}{Z} \sum_{\{n_x\}} e^{-\frac{2\pi^2}{g_o^2} (n_{x+1}-n_x)^2} e^{i\Theta(n_L - n_{-L})} e^{i\Theta'(n_{L'} - n_{-L'})} =$$

$$= \left[\frac{\sum_m e^{-\frac{g_o^2}{8\pi^2}(\Theta+\Theta'+2\pi m)^2}}{\sum_m e^{-\frac{g_o^2}{8\pi^2}(\Theta+2\pi m)^2}} \right]^{2L'} \tag{4.5}$$

(the last step uses Fourier transformation, i.e. the Poisson summation formula (4.3)); this implies

$$\varepsilon(\Theta) \leq \log \sum_m e^{-\frac{g_o^2}{8\pi^2}(\Theta+2\pi m)^2} - \log \sum_m e^{-\frac{g_o^2}{8\pi^2} m^2} \tag{4.6}$$

from which the strict inequality in (2) can be read off.

□

Remarks :

(1) Mack [47] gives a simple argument for confinement of fractional charges in two dimensions that even works for nonabelian groups. We will get this result in subsection d) by a different argument.

(2) Θ-states also exist for nonabelian groups; Θ there has to be an element of the dual of the center (see for instance [66]).

(3) Θ-states in 4-dimensional pure Yang-Mills or Higgs models are hard to find on the lattice because there is no natural lattice analogue of the topological charge density $F \wedge F$. If fermions are present, one can use the angle Θ introduced in Section 1 (see Section 5 and [95]).

(4) Similar results are expected to hold for lattice QED_2 .

We can summarize these results again in a picture:

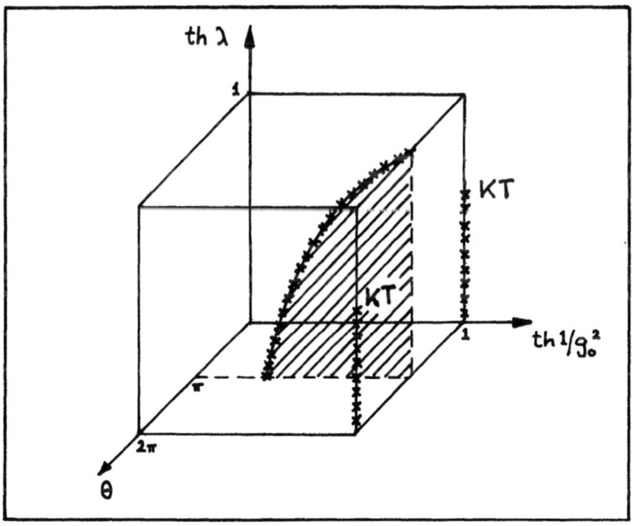

The shaded surface is a phase boundary ending in a critical line; KT stands for the Kosterlitz-Thouless transition point that is also the endpoint of a critical line.

b) The 3-Dimensional Abelian Higgs Model : Phase Structure

The U(1) Higgs model in three dimensions, also known as the Landau-Ginzburg model is of course a very physical one : It is a model of a superconductor.

It is therefore quite interesting that for a Gaussian version of the model the existence of at least two phases, a "normal" and a "superconducting" one, can be established by the use of correlation inequalities [58] and the existence of the Kosterlitz-Thouless transition [60]; in some versions of the model expansions of Section 3 do the job.

The detailed picture depends a little bit on the version of the model studied : "Compact" (i.e. Wilsonian) or Gaussian. The least interesting one has complete breakdown of symmetry (corresponding to the absence of charges smaller than the charge of the Higgs field). This requires the use of a periodic action for the gauge field such as Wilson's or Villain's form. It probably has no phase transitions except at $g^2 = 0$ where it becomes a plane rotator for which a symmetry breaking (Goldstone) transition has been shown to occur in [74]. This model apparently is "always a superconductor".

If we give a shorter period to the Higgs action, i.e. a multiple n of the elementary charge to the Higgs field, we are in the situation of incomplete breakdown of symmetry with $H = \mathbb{Z}_n$ as the unbroken subgroup, and we can use Theorem 3.20' and Quasi-Theorem 3.20. We discover at least two phases : One is "superconducting", with free fractional (with respect to the Higgs field) charges and exponentially clustering $\mathbb{Z}_n$-vortices; the other is "normal" with confined fractional charges and no clustering of vortices.

The most realistic version uses the Gaussian action described in the previous subsection. In $d \geq 3$ and for μ^2 small or zero, this action does not easily permit a high-temperature expansion. This is where correlation inequalities are helping out; the string of arguments is similar as in Subsection a) :

Let us consider the expectation value of a fractionally charged Wilson loop $\langle W_\Theta(C)\rangle$, of rectangular shape, sides L (in the 1-direction) and T (in the 0-direction). As in a) an upper bound is obtained by eliminating A_1 and A_2 and freezing A_o to integer values. Again the different time layers are decoupled and we obtain

$$|\langle W_\Theta(C)\rangle| \leq (\langle e^{i\Theta(n_L - n_o)}\rangle_V)^T \tag{4.7}$$

where the expectation on the right hand side is in the 2-dimensional integer spin (= dual Villain) model with action

$$\frac{1}{2g_o^2} \sum_{\langle xy\rangle} (n_x - n_y)^2 \quad , \quad (n \in \mathbb{Z}) \quad . \tag{4.8}$$

Now in [60] it is shown that $\langle e^{i\Theta(n_L - n_o)} \rangle_L$ behaves like $e^{c \log L}$ provided g^2 is large enough (i.e. $\frac{1}{g_o^2}$ below the Kosterlitz-Thouless transition point). So we obtain

$$\langle W_\Theta(C) \rangle \leq c' \, e^{-cT \log L} \tag{4.9}$$

which gives a logarithmic (i.e. two-dimensional Coulombic) confining potential between fractional electric charges. So there is no screening of these charges; this is the "normal" phase. By arguments "dual" to this one it can be shown that vortices have at most power-like clustering in this phase.

The existence of a superconducting phase with liberated electric charges (perimeter law for W_Θ) and exponentially clustering vortices can be seen by an expansion of the kind used to quasi-prove Quasi-Theorem 3.20.

We close this subsection with some tentative phase diagrams. Note that the correlation inequalities say that the Wilson loop expectation always increases to the right and upward; expectations of vortices go the opposite way. (By these correlation inequalities it is of course also seen that for the Gaussian model

$$\langle W_\Theta(C) \rangle \geq e^{-\tilde{c}\, T \log L}) \quad .$$

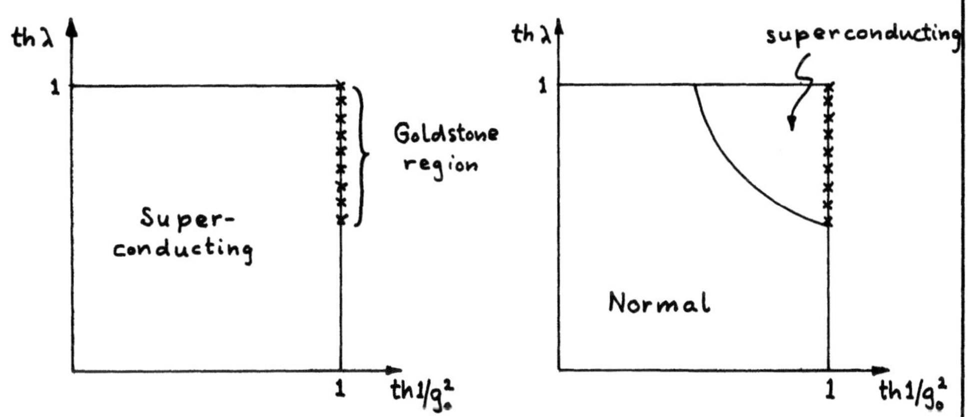

Compact complete breakdown

Compact partial breakdown

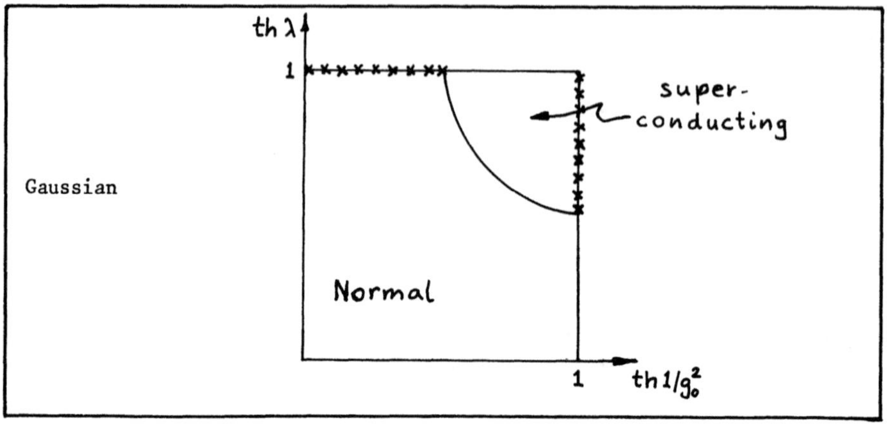

c) Guth's Theorem : Existence of a Nonconfining (Coulombic) Phase in the 4-Dimensional U(1) Model.

Guth [64] gave a clever proof that the pure U(1) lattice gauge model in the Villain form in 4-dimensions behaves more or less like the continuum model provided the coupling is weak enough; in particular there is no confinement. This result is of fundamental importance because it shows that "compact" lattice gauge theories (with fields taking values in the compact group G) are capable of capturing some of the physical content usually ascribed to the continuum models. It also shows that confinement in the sense of Wilson is a subtle phenomenon : 4 is the critical dimension and the delicate dividing line may lie between 4-dimensional abelian and nonabelian models as hoped or anticipated by the physics community.

I will try to give an essentially complete proof of Guth's theorem, but since his paper (in preprint form) had about 50 pages I will have to be brief on some details. Many of the ideas and a lot of the formalism used are already in [77]. The proof presented here differs in many details from Guth's original one.

Remark : Guth's theorem is in some sense an analogous result to the proof of the Thouless-Kosterlitz transition [60]. Fröhlich and Spencer [87] have used the "renorlization group" methods of [60] for an alternate proof. They also prove that even the Z_n lattice gauge model has a "Coulombic" phase for intermediate coupling provided n is large enough, in addition to conventional high and low temperature phases; arguments for this were given already in [61].

First we need some formalism. Throughout we work on a finite lattice Λ ; we will prove a bound independent of Λ . Again we will take Λ to be part of the simple cubic lattice $\mathbb{Z}^4$, but this is in no way essential.

We call functions from the sites (0-cells) of Λ into the integers or reals 0-chains, functions of links (1-cells) 1-chains, functions of plaquettes (2-cells) 2-chains etc. (these are used in analogy to differential forms; some people might therefore prefer to call them cochains). The spaces of chains carry a natural ℓ^2 inner product, so there is a natural identification of chains with cochains.

We define a boundary operator δ from p-chains to (p-1)-chains in the standard way : If ω is a p-chain, $\delta\omega$ assigns for a (p-1)-cell the sum of ω evaluated at all p-cells that contain the (p-1)-cell in question in the right orientation. This is the analogue of the divergence operator.

Example :

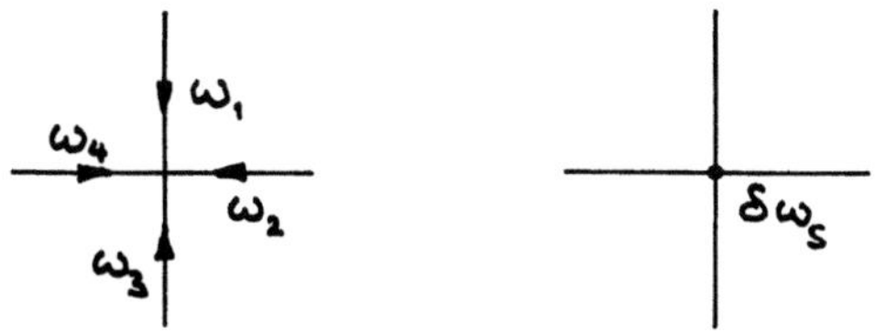

$$\delta\omega_s = \omega_1+\omega_2+\omega_3+\omega_4$$

The ℓ^2 adjoint of δ is denoted by d : it corresponds to the exterior derivative.

Example :

$$d\omega_p = \omega_1+\omega_2+\omega_3+\omega_4$$

The notation used here is different from the standard one of algebraic topology : it reflects, however, the analogy with differential forms.

Note that changing the orientation of a cell flips the sign of the corresponding chain.

By going from Λ to its dual lattice $*\Lambda$ we obtain a duality map $*$ from p-chains into (4-p)-chains corresponding to the Hodge star operator. Two basic facts are

Lemma 4.3 : $\delta^2 = 0 = d^2$.

and

Lemma 4.4. : (Hodge decomposition). Let ω be a p-chain. Then there is a (p+1)-chain α , a (p-1)-chain β and a harmonic p-chain h (i.e. obeying $\delta h = 0 = dh$) such that

$$\omega = \delta\alpha+d\beta+h \tag{4.10}$$

We do not give a proof of these well known and simple algebraic facts (see for instance [98]).

From now on we want to restrict ourselves to topologically trivial lattices Λ for which there are no harmonic chains. This is equivalent to saying that each p-form ω obeying $d\omega = 0$ is of the form $\omega = d\beta$ ("Poincaré's lemma") and we say that Λ has trivial (co)homology. This restriction is not necessary but helps to simplify the arguments. It is true for instance for a rectangular piece of $\mathbb{Z}^4$ with the cells in the geometric boundary omitted ("Dirichlet boundary conditions"), or its dual. Next we define the Laplacean :

Definition 4.5 : $\Delta \equiv d\delta + \delta d$ (Note the sign convention!).

Lemma 4.6 : On p-chains in a 4-dimensional lattice Λ of trivial cohomology we have

$$0 < \Delta < 16 \tag{4.11}$$

Proof : The lower bound (wich can actually be shown to be of the order $|\Lambda|^{-1/2}$) follows from the absence of harmonic chains. The upper bound comes from the fact that the lattice provides an ultraviolet cutoff; it is easy to see in the infinite lattice by Fourier transformation. The finiteness of Λ actually makes Δ smaller because it means replacing the infinite volume Laplacean Δ_∞ by $P\Delta_\infty P$ where P is a projection (cf. [64] for a more detailed discussion).

□

We now turn to the Wilson loop in the U(1) Villain pure gauge model. The Wilson loop $W_n(C)$ can be described by a one-chain j taking an integer value n on links in C and 0 otherwise (in the Villain version it is not sensible to consider Wilson loops with nonintegral charge). j may be thought of as a current running through C . We have

$$\langle W_n(C)\rangle = \frac{1}{Z} \prod_{\langle xy\rangle} \int_{-\pi}^{\pi} d\Theta_{xy} \times$$
$$\times \prod_P \sum_{\ell_P} e^{-\frac{1}{2g_o^2}((d\Theta)_P - 2\pi\ell_P)^2} \prod_{\langle xy\rangle} e^{i(j_{xy},\Theta_{xy})} \tag{4.12}$$

From here on Z will always denote the appropriate normalization factor and may change its meaning from line to line.

Theorem 4.7 : (Guth [64]) : For g_o^2 small enough (actually $g_o^2 < 0.168$) there is a function $\bar{g}(g_o)$ such that

$$\langle W_n(C)\rangle \geq \exp(-\frac{\bar{g}(g_o)^2}{2}(j,\Delta^{-1}j))$$

and

$$\lim_{g\to 0} \frac{\bar{g}(g_o)^2}{g^2} = 1$$

Remark : This implies a perimeter law for the Wilson loop.

To prove this theorem one first rewrites $\langle W_n(C)\rangle$ in a form given by Banks, Kogut and Myerson [65] and closely related to the description given by Glimm and Jaffe in [78].

Lemma 4.8 : $\langle W_n(C)\rangle = A_o(C)E(C)$ where

$$A_o(C) = \exp(-\frac{g_o^2}{2}(j,\Delta^{-1}j))$$

$$E(C) = \frac{1}{Z} \sum_{\substack{m_3 \\ dm_3=0}} \exp(-\frac{1}{2g_o^2}(m_3,\Delta^{-1}m_3)) \exp(2\pi i(m_3,\Delta^{-1}dG_2))$$

where G_2 is an integer valued 2-chain obeying $\delta G_2 = j$ (think of a surface bordered by C). The sum is over integer valued 3-chains m_3 .

Cor. 4.9 : $\langle W_n(C)\rangle \leq \exp(-\frac{g_o^2}{2}(j,\Delta^{-1}j))$.

Proof : Trivial. □

Remark : Glimm and Jaffe [78] had proven this for Wilson's action.

Proof of the lemma : We first fix the gauge for Θ in (4.12) by selecting a maximal tree T of the lattice Λ on which Θ can be put equal to zero. We write $\int'_{\Theta_1} d\Theta_1$ for the resulting integral over 1-chains Θ_1 with values in $[-\pi,\pi]$ and vanishing on T . Thus

$$\langle W_n(C)\rangle = \frac{1}{Z} \sum_{\ell_2} \int'_{\Theta_1} d\Theta_1\, e^{i(j,\Theta_1)} \exp(-\frac{1}{2g_o^2}\| d\Theta_1 - 2\pi\ell_2\|^2) \tag{4.13}$$

If we denote $d\ell_2$ by m_3 we can write

$$\ell_2 = \ell_2[m_3] + d\ell_1$$

where $\ell_2[m_3]$ is a particular integer valued solution of $m_3 = d\ell_2$. We may then replace $\sum_{\ell_2}$ by $\sum_{\substack{m_3 \\ dm_3=0}} \sum'_{\ell_1}$ where $\sum'_{\ell_1}$ means that ℓ_1 has been set equal to zero on the maximal tree T (to make ℓ_1 uniquely defined by $d\ell_1 = \ell_2 - \ell_2[m_3]$). This gives

$$\langle W_n(C)\rangle = \frac{1}{Z} \sum_{\substack{m_3 \\ dm_3=0}} \sum'_{\ell_1} \int'_{\Theta_1} d\Theta_1\, e^{i(j,\Theta_1)} \times \exp\left(-\frac{1}{2g_o^2}\| d\Theta_1 - 2\pi\ell_2[m_3] + 2\pi d\ell_1 \|^2\right) \tag{4.14}$$

By the Hodge decomposition and the absence of harmonics $1 = \delta\Delta^{-1}d + d\Delta^{-1}\delta$ and we have

$$\ell_2[m_3] = d\Delta^{-1}\delta\ell_2[m_3] + \delta\Delta^{-1}m_3 \tag{4.15}$$

hence

$$\begin{aligned}\langle W_n(C)\rangle &= \frac{1}{Z} \sum_{\substack{m_3 \\ dm_3=0}} \sum'_{\ell_1} \int'_{\Theta_1} d\Theta_1\, e^{i(j,\Theta_1+2\pi\ell_1)} \\ &\quad \times \exp\left(-\frac{1}{2g_o^2}\| \delta\Delta^{-1}m_3 \|^2\right) \exp\left(-\frac{1}{2g_o^2}\| d(\Theta_1 + 2\pi\ell_1 + 2\pi\Delta^{-1}\delta\ell_2[m_3]) \|^2\right) \\ &= \frac{1}{Z} \sum_{\substack{m_3 \\ dm_3=0}} \int'_{\alpha_1} d\alpha_1\, e^{i(j,\alpha_1)} \exp\left(-\frac{1}{2g_o^2}(m_3, \Delta^{-1}m_3)\right) \\ &\quad \times \exp\left(-\frac{1}{2g_o^2}\| d\alpha_1 \|^2\right) e^{-i(j,\delta\Delta^{-1}2\pi\ell_2[m_3])}\end{aligned} \tag{4.16}$$

where we put $\alpha_1 = \Theta_1 + 2\pi\ell_1 + 2\pi\delta\Delta^{-1}\ell_2[m_3]$; $\int'_{\alpha_1} d\alpha_1$ is now over all real valued 1-chains vanishing on T . This Gaussian integral now can be computed and gives

$$\langle W_n(C)\rangle = \exp\left(-\frac{g_o^2}{2}(j,\Delta^{-1}j)\right)$$

$$\times \sum_{\substack{m_3 \\ dm_3=0}} \exp\left(-\frac{1}{2g_o^2}(m_3,\Delta^{-1}m_3)\right) e^{-2\pi i(j,\delta\Delta^{-1}\ell_2[m_3])} \tag{4.17}$$

By the definition of G_2

$$(j_1,\delta\Delta^{-1}\ell_2[m_3]) = (G_2,d\Delta^{-1}\delta\ell_2[m_3]) = (G_2,\ell_2[m_3]) - (G_2,\delta\Delta^{-1}m_3) \tag{4.18}$$

The first term is an integer and does not contribute in the exponential, so we finally obtain

$$\langle W_n(C)\rangle = A_o(C) \sum_{\substack{m_3 \\ dm_3=0}} \exp\left(-\frac{1}{2g_o^2}(m_3,\Delta^{-1}m_3)\right) \exp(2\pi i(m_3,\Delta^{-1}dG_2)) \tag{4.19}$$

as asserted. □

Corollary 4.10 : Let $a_1 \equiv *\Delta^{-1}dG_2$. Then

$$E(C) = \frac{1}{Z} \sum_{\substack{m_1 \\ dm_1=0}} e^{-\frac{1}{2g_0^2}(m_1,\Delta^{-1}m_1)+2\pi i(m_1,a_1)} \tag{4.20}$$

Proof : Trivial.

We now bound $E(C)$ from below by a simple correlation inequality (closely related to the inequalities of Fröhlich and Park [79]) :

Lemma 4.11 : Let $0 < A_1 \leq A_2$; A_1, A_2 operators on p-chains. Then

$$\frac{1}{Z_1} \sum_{m_p} e^{-(m_p,A_1m_p)+i(a_p,m_p)} \leq \frac{1}{Z_2} \sum_{m_p} e^{-(m_p,A_2m_p)+i(a_p,m_p)} .$$

Corollary 4.12 :

$$E(C) \geq \frac{1}{Z} \sum_{\substack{m_1 \\ \delta m_1=0}} e^{-\frac{1}{2}\frac{1}{16g_o^2}(m_1,m_1)+2\pi i(m_1,a_1)} \equiv F(C) \tag{4.12}$$

Proof of the corollary : This becomes obvious if we replace the constraint $\delta m_1 = 0$ by a suitable limit and use Lemma 4.6.

□

Proof of the lemma : Let

$$F(\lambda) \equiv \frac{1}{Z(\lambda)} \sum_{m_P} e^{-(m_P, A(\lambda) m_P) + i(a_P, m_P)}$$

where $A(\lambda) = \lambda A_2 + (1-\lambda) A_1$. We claim that $F'(\lambda) \geq 0$:

$$F'(\lambda) = \langle (m_P (A_1 - A_2) m_P) e^{i(a_P, m_P)} \rangle_\lambda -$$

$$- \langle m_P, (A_1 - A_2) m_P \rangle_\lambda \langle e^{i(a_P, m_P)} \rangle_\lambda =$$

$$= \frac{1}{2} \frac{1}{Z(\lambda)^2} \sum_{\substack{m_P \\ m'_P}} e^{-(m_P, A(\lambda) m_P) - (m'_P, A(\lambda) m'_P)} \times$$

$$\times [(m_P, (A_1 - A_2) m_P) - (m'_P, (A_1 - A_2) m'_P)] \times$$

$$\times (e^{i(a_P, m_P)} - e^{i(a_P, m'_P)}) .$$

Introducing

$$m_\pm \equiv m_P \pm m'_P$$

we obtain

$$F'(\lambda) = \frac{i}{Z(\lambda)^2} \sum_{m_+, m_-} e^{-\frac{1}{2}(m_+, A(\lambda) m_+) - \frac{1}{2}(m_-, A(\lambda) m_-)} \times$$

$$\times (m_+, A_1 - A_2) m_-) e^{\frac{i}{2}(a_P, m_+)} \sin \frac{1}{2}(a_P, m_-)$$

Since $F'(\lambda)$ is invariant under $m_+ \to -m_+, m_- \to -m_-$, we have

$$F'(\lambda) = \frac{1}{Z(\lambda)^2} \sum_{m_+, m_-} e^{-\frac{1}{2}(m_+, A(\lambda) m_+) - \frac{1}{2}(m_-, A(\lambda) m_-)} \times$$

$$\times \sin \frac{1}{2}(a_P, m_+) \sin \frac{1}{2}(a_P, m_-) \times$$

$$\times (m_+, (A_2 - A_1) m_-) \geq 0 .$$

The last inequality is true because $F'(\lambda)$ is of the form $(\psi, (A_2 - A_1)\psi)$ and

and $A_2 \geq A_1$. □

$F(C)$, defined in (4.21), can be rewritten as the expectation value of a disorder variable (a very nonlocal one) in a plane rotator model at high temperature (see [64]). It is, however, simpler to obtain a cluster expansion for $\log F(C)$ from (4.21) directly by using the low temperature expansion in "defects" (which here should be interpreted as magnetic currents).

So let us put our machine constructed in Section 3 in gear to grind out a lower bound on $F(C)$: We have

$$\log F(C) = \sum_X (z_{a_1}^X - z_o^X) \frac{a(X)}{X!} \tag{4.22}$$

(cf (3.35)). As polymers γ we take here the possible supports of 1-chains m_1 with $\delta m_1 = 0$, i.e. connected networks of lattice lines without endpoints. The activities are

$$z_{a_1}(\gamma) = \sum_{\substack{m_1 \neq 0 \\ \text{on } \gamma; \\ \delta m_1 = 0}} e^{-\frac{1}{32 g_o^2} \|m_1\|^2 + 2\pi i (a_1, m_1)} \tag{4.23}$$

Each γ can be written as a union of closed loops :

$$\gamma = \bigcup_i C_i \tag{4.24}$$

and for each closed loop C_i we can find a sheet S_i bordered by C_i . Let

$$S(\gamma) = \bigcup_i S_i \quad . \tag{4.25}$$

On S , considered as a sublattice, we can find a 2-chain $m_2[m_1]$ obeying $\delta m_2[m_1] = 0$ for any m_1 on γ with $\delta m_1 = 0$. This can be used to rewrite (4.23) in a more "gauge invariant" form. Recall that

$$a_1 = *\Delta^{-1} dG_2 \quad .$$

Thus

$$da_1 = *\delta\Delta^{-1} dG_2 = *G_2 - *d\Delta^{-1}\delta G_2 = *G_2 - *d\Delta^{-1} j \quad . \tag{4.26}$$

Defining the "electromagnetic field" F_2 generated by the current j by

$$F_2 \equiv d\Delta^{-1}j \tag{4.27}$$

we have $da_1 = *G_2 - *F_2$ and

$$z_{a_1}(\gamma) = \sum_{\substack{m_1 \neq 0 \\ \text{on } \gamma; \\ \delta m_1 = 0}} e^{-\frac{1}{32g_o^2}\|m_1\|^2 - 2\pi i(*F_2, m_2[m_1])} \tag{4.28}$$

($*G_2$ drops out because it is integer valued).

A crucial fact to note is that $S(\gamma)$ can be defined in such a way that we have the "isoperimetric inequality"

$$|S(\gamma)| < \frac{1}{16}|\gamma|^2 \tag{4.29}$$

(4 being "the lattice value of π"). This gives us a crucial estimate :

<u>Lemma 4.14</u> : There is a constant K such that

$$|z_{a_1}(\gamma) - z_o(\gamma)| \leq K|z_o(\gamma)| \; |\gamma|^2 \, \|F_2\|^2_{S(\gamma)} \tag{4.30}$$

where $\|F_2\|^2_{S(\gamma)} \equiv \sum_{p \in S(\gamma)} |F_2(p)|^2$; this is the "electromagnetic energy of the field of j inside $S(\gamma)$ ".

<u>Proof</u> :

$$|z_{a_1}(\gamma) - z_o(\gamma)| \leq 2\pi^2 \sum_{\substack{m_1 \neq 0 \\ \text{on } \gamma; \\ \delta m_1 = 0}} e^{-\frac{1}{32g_o^2}\|m_1\|^2} |(m_2, *F_2)|^2 \; . \tag{4.31}$$

Now

$$|(m_2, *F_2)|^2 \leq \|m_2\|^2 \, \|F_2\|^2_{S(\gamma)}$$

$$\leq \sup_{m_2 \upharpoonright S(\gamma)} |m_2|^2 \, \|F_2\|^2_{S(\gamma)} |S(\gamma)|$$

$$\leq \|m_1\|^2 \, \|F_2\|^2_{S(\gamma)} \frac{1}{16}|\gamma|^2 \; .$$

Using

$$\sum_{\substack{m_1 \neq 0 \\ \text{on } \gamma; \\ \delta m_1 = 0}} \|m_1\|^2 e^{-\frac{1}{32g_o^2}\|m_1\|^2} \leq \tilde{K} \sum_{\substack{m_1 \neq 0 \\ \text{on } \gamma; \\ \delta m_1 = 0}} e^{-\frac{1}{32g_o^2}\|m_1\|^2} \tag{4.32}$$

for g_o^2 small, (4.30) follows (a possible choice is $g_o^2 < 1/(32 \log 8)$, $\tilde{K} = \sum_{m \neq 0} m^2 8^{-m^2} / \sum_{m \neq 0} 8^{-m^2}$) .

Corollary 4.15 :

$$|z_{a_1}^X - z_o^X| \leq \sum_\gamma z_o^X |\gamma|^2 K \|F_2\|_{S(\gamma)}^2 X(\gamma) . \tag{4.33}$$

Proof : This follows from the lemma by telescoping the difference and using $|z_{a_1}(\gamma)| \leq z_o(\gamma)$.

□

The next step is to sum (4.33) over all translates of a given X_o :

Lemma 4.16 :

$$\sum_{\substack{X \text{ translate} \\ \text{of } X_o}} |z_{a_1}^X - z_o^X| \leq K\, z_o^{X_o} \sum_\gamma |\gamma|^3 X_o(\gamma) \|F_2\|_\Lambda^2 \tag{4.34}$$

Proof :

$$\sum_{\substack{\gamma \text{ translate} \\ \text{of } \gamma_o}} \|F_2\|_\gamma^2 \leq |\gamma_o| \, \|F_2\|^2 .$$

□

Now we only have to note $\sum_\gamma |\gamma|^3 X_o(\gamma) \leq (\sum_\gamma |\gamma| X_o(\gamma))^3 \leq a_\varepsilon e^{\varepsilon \Sigma |\gamma| X_o(\gamma)}$ for any $\varepsilon > 0$ provided a_ε is chosen appropriately; we can then conclude

$$\sum_X |z_{a_1}^X - z^X| a(X) \leq K \|F_2\|_\Lambda^2 a_\varepsilon \sum_{X \in 0} z_o^X e^{\varepsilon \sum_\gamma |\gamma| X(\gamma)} a(X)$$

$$\leq K' \|F_2\|_\Lambda^2 e^{-\frac{1}{32g_o^2}}$$

with some constant K' .

Since $\|F_2\|_\Lambda^2 = (j, \Delta^{-1} j)_\Lambda$ we have proven

$$\log F(C) \geq -K' e^{-\frac{1}{32g_o^2}(j,\Delta^{-1}j)_\Lambda} \tag{4.35}$$

which completes the proof of Theorem 4.7. □

For completeness we should note the following fact concerning the disorder parameter in the U(1) Villain gauge model :

Theorem 4.7' : For any value of g_o^2

$$\langle D_\Theta(C)\rangle \geq \exp[-\frac{\Theta^2}{2g_o^2}(j_1,\Delta^{-1}j_1)]$$

where j_1 is an integer valued one-chain on the dual lattice given by

$$j_1(\langle xy\rangle) = \begin{cases} 1 : \langle xy\rangle \in C \\ 0 \quad \text{otherwise} \end{cases}$$

Proof : By definition

$$\langle D_\Theta(C)\rangle = \frac{1}{Z} \prod_{\langle xy\rangle} \int_{-\pi}^{\pi} d\Theta_{xy} \prod_P \sum_{\ell_P} e^{-\frac{1}{2g_o^2}((d\Theta)_P - 2\pi\ell_P - \Theta G_P)^2}$$

where $G_P \equiv G_2(P)$ and G_2 is an integer valued 2-chain obeying $dG_2 = {*j_1}$. By a duality transformation (= Fourier transformation) we get

$$\langle D_\Theta(C)\rangle = \frac{1}{Z} \sum_{\substack{m_2 \\ \delta m_2 = 0}} e^{-\frac{g_o^2}{2}\|m_2\|^2 - i\Theta(m_2,G_2)} .$$

Now we use (a slight extension of) the correlation inequality of Lemma 4.11 to replace the sum over m_2 by a integral over real valued two-chains ω_2 :

$$\langle D_0(C)\rangle > \frac{1}{Z} \int_{\delta\omega_2 = 0} e^{-\frac{g_o^2}{2}\|\omega_2\|^2 - i\Theta(\omega_2,G_2)} =$$

$$= \exp[-\frac{\Theta^2}{2g_o^2}(G_2, \delta\Delta^{-1}d\, G_2)] =$$

$$= e^{-\frac{\Theta^2}{2g_o^2}(j_1,\Delta^{-1}j_1)} .$$

□

So here we have a situation where both loops have perimeter decay. This seems to be characteristic for a "Coulombic" phase.

Remark 1 : One could play exactly the same game for the plane rotator in two dimensions, trying to prove the Kosterlitz-Thouless transition. Instead of the Wilson loop one would consider a two-point function

$$S_n(x-y) \equiv \langle e^{in(\Theta_x - \Theta_y)} \rangle \quad . \tag{4.36}$$

In this case one has

$$S_n(x-y) \geq e^{-\frac{1}{2} g_o^2 V_c(x-y)} F(x-y) \quad , \tag{4.37}$$

where V_c is the lattice Coulomb potential and

$$F(x-y) = \frac{1}{Z} \sum_{m_o} e^{-\frac{1}{8g_o^2}(m_o,m_o)+2\pi i(m_o,a_o)} \tag{4.38}$$

and again

$$e^{2\pi i(m_o,a_o)} = e^{-2\pi i(m_1,{}^*E_1)} \tag{4.39}$$

with $m_1[m_o]$ such that $m_1 = m_o$ and E_1 is the electric field of a change $+n$ at x and a charge $-n$ at y (defined in such a way that it vanishes outside the charge pair).

But if one cluster expands one encounters a problem : The analogue of (4.29) is not true because the length of a string is not controlled by the number of its endpoints. This is why the more sophisticated treatment of Fröhlich and Spencer [60] is necessary which does not throw out the Coulomb interaction by a correlation inequality and uses instead electrostatic methods, together with renormalization group ideas to bound the activities.

Remark 2 : By analogous methods the existence of long range order (spontaneous symmetry breaking) for plane rotator models in $d \geq 3$ can be proven (exercise!).

In some ways this method is stronger than the infrared bounds of ref. [74] : It displays clearly the dominating spin waves (Goldstone modes) and it does not depend on any regularity of the lattice.

d) SU(n) Confines if $\mathbb{Z}_n$ does.

This type of result that shows that SU(n) gauge theories go "in the right direction" from abelian $\mathbb{Z}_n$ theories was first proven by Mack and Petkova [67] for a modified type of model that was invented for this purpose and soon afterwards by Fröhlich [68] for the standard (Wilson) pure Yang-Mills or Yang-Mills-Higgs models. His result also implies the one of Mack [47] mentioned earlier, concerning the two-dimensional models.

We consider a Yang-Mills-Higgs model in which the Higgs representation U_H represents some subgroup Z^o of the center $\mathbb{Z}_n$ of SU(n) trivially (in order to have a chance for confinement of "fractional charges"); the pure Yang-Mills model can be considered a special case of this.

Integrating out the Higgs field we obtain the partition functions

$$Z(\{g_{xy}\})$$

studied in Section 2. Here it should be noted that they are really functions of gauge fields with values in $SU(n)/Z^o$. Also note the simple identity

$$\int_G dg\, F(g) = \int_G dg \int_{Z^o} d\omega\, F(g\omega) \tag{4.41}$$

valid for any compact group G and a central subgroup Z^o.

We now consider a Wilson loop $W_\tau(C)$ belonging to a representation that represents Z^o nontrivially (i.e. corresponding to "fractional charges"). We compute

$$\begin{aligned}\langle W_\tau(C)\rangle_\Lambda &= Z_\Lambda^{-1} \int \chi_\tau(g_c) e^{-S_{Y.M.}(\{g_{xy}\})} Z(\{g_{xy}\}) \prod_{\langle xy\rangle} dg_{xy} = \\ &= Z_\Lambda^{-1} \int \prod_{\langle xy\rangle} dg_{xy}\, Z(\{g_{xy}\}) \chi_\tau(g_c) \times \\ &\times \prod_{\langle xy\rangle} \int_{Z^o} d\gamma_{xy} \chi_\tau(\gamma_c) e^{-S_{Y.M.}(\{\gamma_{xy} g_{xy}\})}\end{aligned} \tag{4.42}$$

where we employed (4.41). Inserting the concrete form of the Wilson action we obtain (using the fundamental representation of SU(n) in the action) :

$$\langle W_\tau(C)\rangle_\Lambda = \frac{1}{Z_\Lambda}\int \prod_{\langle xy\rangle} dg_{xy}\; Z(\{g_{xy}\})\chi_\tau(g_c) \times$$

$$\times \prod_{\langle xy\rangle} \int_{Z^o} d\gamma_{xy} \frac{\chi_\tau(\gamma_C)}{\chi_\tau(\mathbb{1})}\; e^{-\frac{1}{g_o^2}\sum_P \mathrm{Re}\,\chi(g_{\partial P})\frac{\chi(\gamma_{\partial P})}{\chi(\mathbb{1})}} \tag{4.43}$$

This can be viewed as a Z^o-gauge theory with fluctuating coupling constants, a point of view that is often useful (cf. [67]).

We now use the fact that

$$\chi(\gamma) = e^{i\theta_\gamma}\chi(\mathbb{1}) \tag{4.44}$$

$$\chi_\tau(\gamma) = e^{iq_\tau\theta_\gamma}\chi_\tau(\mathbb{1}) \tag{4.45}$$

for some integer q_τ and some angles θ_γ, to rewrite the essential piece of the action in (4.43) :

$$\begin{aligned} &\mathrm{Re}\;\chi(g_{\partial P})\chi(\gamma_{\partial P})/\chi(\mathbb{1}) \\ &= \mathrm{Re}\;\chi(g_{\partial P})\cos\theta_{\partial P} - \mathrm{Im}\;\chi(g_{\partial P})\sin\theta_{\partial P} \\ &\equiv J_P\cos\theta_{\partial P} + K_P\cos(\theta_{\partial P}+\tfrac{\pi}{2}) \end{aligned} \tag{4.46}$$

where we simplified the notation $\theta_{\gamma_{\partial P}}$ to $\theta_{\partial P}$.

We define a $\{g_{xy}\}$ -dependent probability measure

$$d\mu(\{\theta_{xy}\})_{J,K} \equiv \frac{1}{\widetilde{Z}(\{g_{xy}\})}\prod_{\langle xy\rangle} d\lambda(\theta_{xy}) \times \prod_P e^{-\frac{1}{g_o^2}(J_P\cos\theta_{\partial P}+K_P\cos(\theta_{\partial P}+\frac{\pi}{2})} \tag{4.47}$$

where $\widetilde{Z}$ is the obvious normalization and $d\lambda(\theta)$ is the Haar measure of Z^o parametrized by θ .

The goal is now to get rid of the fluctuating coupling constants J_p, K_p by some correlation inequality. The relevant inequality has been provided by Messager, Miracle-Solé and Pfister [69], and says that for $q \in \mathbb{Z}$, $\alpha \in \mathbb{R}$

$$\pm \int \cos(q\theta_c+\alpha)d\mu_{J,K} \leq$$

$$\leq \frac{1}{Z'} \int \cos(q\theta_c) e^{-\frac{1}{g'^2}\sum_P \cos\theta_{\partial P}} d\lambda(\{\theta_{xy}\}) \tag{4.48}$$

provided

$$\frac{1}{g_o^2}(|J_P|+|K_P|) \leq \frac{1}{g'^2} \tag{4.49}$$

(actually to apply [69] one has to write $d\lambda(\theta)$ as weak limit of $\frac{1}{z(\mu)} e^{\mu \cos m\theta} d\theta$ for suitable m and $\mu \to \infty$) .

Now (4.49) is true provided

$$\frac{1}{g'^2} \geq \frac{2d}{g_o^2} \quad , \quad d = \chi(\mathbb{1}) \tag{4.50}$$

Inserting (4.48) in (4.43), using (4.46) and (4.47) gives

$$|\langle W_\tau(C)\rangle^{SU(n)}_{\Lambda,g^2}| \leq 2\chi(\mathbb{1})\langle\cos q\theta_c\rangle^{\mathbb{Z}_n}_{\Lambda,g^2/2d} \tag{4.51}$$

We summarize what we have proven :

Theorem 4.17 : In a SU(n) lattice Yang-Mills-Higgs theory of coupling constant g , where the Higgs representation acts trivially on a subgroup Z^o contained in the center of SU(n) , a fractionally charged Wilson loop has area decay, provided the same Wilson loop has area decay in the pure Z^o Yang-Mills theory of coupling constant $\frac{g}{\sqrt{2\chi(\mathbb{1})}}$.

Corollary 4.18 : In two dimensions fractional charges are always confined.

e) The Interplay of Electric and Magnetic Properties in the Confinement Problem.

The common view nowadays is that nonabelian gauge theories confine "electric" charges because they behave like magnetic superconductors and squeeze the electric flux the way a superconductor squeezes the magnetic flux. This qualitative picture is supported by some rigorous results which we will discuss here : A confinement criterion due to Mack and Petkova [70] shows that "vortex condensation" in connection with spreading of magnetic flux leads to confinement. Mack and Petkova [67] also showed that introducing a constraint that eliminates monopoles of some kind and makes spreading of magnetic flux more difficult produces confinement of external monopoles at weak coupling; we also show that it leads to perimeter decay of a certain type of Wilson loop. Finally there is a general electric-magnetic duality relation due to 't Hooft [76] that shows that squeezing of flux may occur either for the electric or the magnetic kind but not for both - this is, by the way, in agreement with the expansion results of Section 3.

The criterion of Mack and Petkova [70] is inspired by the work of Dobrushin and Shlosman [71] on two-dimensional spin systems with continuous symmetry. They showed that the Mermin-Wagner theorem [72] on the absence of spontaneous magnetization in these spin systems can be rederived if one uses the old intuitive idea that long Bloch walls (Peierls contours) can be made to cost little free energy by making them thick and making the spins change slowly.

In lattice gauge theories the analogue of Peierls contours are of course the defects we have been talking about so much. If the gauge group is continuous our low temperature expansions do not work because the defects get fuzzy. This fuzziness may lead to an area law for the Wilson loop even for small g_o (provided the dimension is not too large, i.e. ≤ 4).

As possible vortex containers we consider finite regions Λ of our infinite lattice $\mathbb{Z}^d$ with a "toroidal" topology, i.e. we assume that the boundary $\partial\Lambda$ is homeomorphic to $S^{d-2} \times S^1$. Consider a corresponding partition function $Z_\Lambda(g_{\partial\Lambda})$ for a pure lattice Yang-Mills theory with given values of the gauge field on the boundary. We now want to give a "twist" to these boundary conditions by acting on them with a "singular gauge transformation" (see Section 2) in such a way that all the holonomy operators (= Wilson loops) corresponding to loops that are running around Λ in the "thin" direction (see picture) are multiplied by an element ω of the center of our gauge group G .

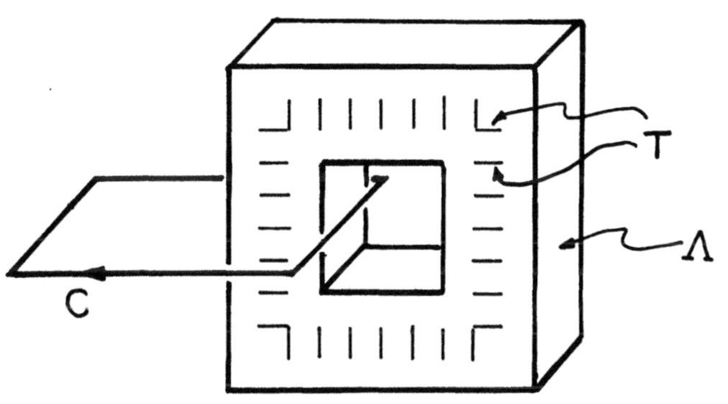

A 3-d vortex container Λ linked to a loop C

This can be achieved by picking a (d-2)-cycle $*T$ on $\partial\Lambda$ (recall that $\partial\Lambda$ has a factor S^{d-2}) and multiplying all gauge fields living on its dual T by ω.

The corresponding partition function we denote by $Z_{\Lambda,\omega}(g_{\partial\Lambda})$. We define a "vortex probability" :

Definition 4.19 :

$$p_{\Lambda,g_{\partial\Lambda}}(\omega) \equiv Z_{\Lambda,\omega}(g_{\partial\Lambda})\left(\int_Z d\omega'\, Z_{\Lambda,\omega'}(g_{\partial\Lambda})\right)^{-1} \tag{4.52}$$

where $d\omega$ is the normalized Haar measure on the center Z of G (usually discrete); note that $Z_{\Lambda,\omega}(g_{\partial\Lambda}) > 0$.

Also important is its Fourier transform

$$\hat{p}_{\Lambda,g_{\partial\Lambda}}(q) \equiv \int_Z d\omega\chi_q(\omega)p_{\Lambda,g_{\partial\Lambda}}(\omega) \tag{4.53}$$

(q runs through the dual of Z).

Now pick any N vortex containers that are disjoint and are linked to a given closed loop C as indicated in the picture. Then we have :

Theorem 4.20 [70]:

$$|\langle W_\tau(C)\rangle| \leq \chi_\tau(\mathbb{1}) \prod_{i=1}^{N} \sup_{g_{\partial\Lambda_i}} |\hat{p}_{\Lambda_i,g_{\partial\Lambda_i}}(q_\tau)| \tag{4.54}$$

where q_τ is the representation induced by τ on the center of G (assumed to be irreducible).

Proof : Let $\Lambda^c \equiv \Lambda^d \setminus \bigcup_{i=1}^{N} \overset{o}{\Lambda}_i$ where $\overset{o}{\Lambda}_i$ is the interior of Λ_i (here considered as a set of links). Then

$$\langle W_\tau(C)\rangle_\Lambda = \langle \chi_\tau(g_c) \prod_{i=1}^{N} Z_{\Lambda_i,\mathbb{1}}(g_{\partial\Lambda_i})\rangle_{\Lambda^c}\, Z_{\Lambda^c}/Z_\Lambda \tag{4.55}$$

We now make a change of variables on the right hand side of (4.55) corresponding to a singular gauge transformation (i.e. leaving all plaquette variables unchanged) in Λ^c, and giving a twist ω_i to the boundary conditions $g_{\partial\Lambda_i}$ $(i = 1,\ldots,N)$; it is easy to see that this is possible. We see

$$\langle W_\tau(C)\rangle_\Lambda = \left\langle \frac{\chi_\tau\left(\prod_{i=1}^{N}\omega_i\right)}{\chi_\tau(\mathbb{1})}\,\chi_\tau(g_C)\prod_{i=1}^{N} Z_{\Lambda_i,\omega_i}(g_{\partial\Lambda_i})\right\rangle_{\Lambda^c} Z_{\Lambda^c}/Z_\Lambda$$

Averaging over $\omega_1,\ldots,\omega_N$ we obtain

$$\langle W_\tau(C)\rangle_\Lambda = \langle \prod_{i=1}^{N} [\hat{p}_{\Lambda_i,g_{\partial\Lambda_i}}(q_\tau)\int d\omega_i Z_{\Lambda_i,\omega_i}(g_{\partial\Lambda_i})]\chi_\tau(g_C)\rangle_{\Lambda^c}\, Z_{\Lambda^c}/Z_\Lambda \tag{4.56}$$

Using $|\chi_\tau(g_C)| \leq \chi_\tau(\mathbb{1})$ and

$$\langle \prod_{i=1}^{N} Z_{\Lambda_i,\mathbb{1}}(g_{\partial\Lambda_i})\rangle_{\Lambda^c} = Z_\Lambda/Z_{\Lambda^c}$$

we obtain the assertion from (4.56). □

In order to interprete this theorem we show how it allows to deduce confinement, i.e. almost an area law for the Wilson loop, from a property of the free energy of vortices (that has not been proven for weak coupling !).

Let us assume that the free energy per unit length (in $d = 3$) or unit surface area (in $d = 4$) of a vortex container becomes rapidly independent of twists in the boundary condition if the thickness increases; more precisely, assume

$$\left|\log \frac{Z_{\Lambda,\omega}(g_{\partial\Lambda})}{Z_{\Lambda,\mathbb{1}}(g_{\partial\Lambda})}\right| \leq |T|e^{-md} \tag{4.57}$$

where d is the "thickness" of Λ . Then d can be chosen large enough to obey $|T|e^{-md} \leq \xi$ with ξ sufficiently small.

For simplicity let us assume that $G = SU(2)$. Then there is only one nontrivial character χ_q and for this q we have

$$\hat{p}_{\Lambda,g_{\partial\Lambda}}(q) = \frac{Z_{\Lambda,\mathbb{1}}(g_{\partial\Lambda})-Z_{\Lambda,-\mathbb{1}}(g_{\partial\Lambda})}{Z_{\Lambda,\mathbb{1}}(g_{\partial\Lambda})+Z_{\Lambda,-\mathbb{1}}(g_{\partial\Lambda})} =$$

$$= \frac{1-Z_{\Lambda,-\mathbb{1}}(g_{\partial\Lambda})/Z_{\Lambda,\mathbb{1}}(g_{\partial\Lambda})}{1+Z_{\Lambda,-\mathbb{1}}(g_{\partial\Lambda})/Z_{\Lambda,\mathbb{1}}(g_{\partial\Lambda})}$$

By the assumption (4.57) we then get

$$|\hat{p}_{\Lambda,g_{\partial\Lambda}}(q)| \le \frac{e^{\xi}-1}{e^{-\xi}-1} \le (e^{\xi}-1)(e^{-\xi}+1) = 2 \sinh \xi \quad .$$

If the Wilson loop C is large enough to accommodate N vortex containers of the required thickness, the theorem says then

$$|\langle W_\tau(C)\rangle| \le 2(2 \sinh \xi)^N$$

(if τ is the fundamental representation of $SU(2)$) .

The rest is a packing problem. A little thought shows that a rectangular loop of sides L and T can accomodate of the order $\frac{LT}{(\log L)^2}$ vortex containers with length $|T_i|$ and thickness d_i obeying $|T_i|e^{-md_i} \le \xi$ for all i . This does not quite give the area law, but it corresponds to a potential

$$V(L) = \mathcal{O}\left(\frac{L}{(\log L)^2}\right)$$

which is confining.

It should be noted that the assumption (4.57) can be proven for strong coupling by the cluster expansion quite routinely; this has been done in [73].

In [16] Mack and Petkova study a pure lattice Yang-Mills theory with gauge group $SU(2)$ that is modified by the harmless looking constraint

$$\prod_{P\in\partial C} \chi(g_{\partial P}) \ge 0 \tag{4.58}$$

for all 3-cells (cubes) C , where χ is the character appearing in the action - here χ is the character of the fundamental representation of $SU(2)$. (4.58) seems to be automatically fulfilled for weak coupling because $g_{\partial P}$ wants to be near $\mathbb{1}$ there; the formal continuum limit is not affected by (4.58). But Mack and Petkova prove that for $d \ge 3$ and weak coupling the disorder variables show a drastically different behavior due to (4.58); for instance in $d = 4$ the 't Hooft loop shows area decay. This is an indication that confinement might break down and we will show directly that there is a Wilson loop with perimeter decay in the modified model.

This is a warning sign that shows that it is not irrelevant which lattice approximation is chosen and it is not enough to check the formal continuum limit (we saw already how important the choice of the character occurring in the action is). The modified model probably has a critical point at nonzero coupling and this seems to be in conflict with asymptotic freedom.

(4.58) has the form of a Bianchi identity for the Z_2 variables $\operatorname{sgn} \chi(g_{\partial P})$. It prevents thin defects defined by $\chi(g_{\partial P}) < 0$ from branching and thereby spreading.

(4.58) allows to derive the $\mathbb{Z}_2$ valued 2-chain ω_2 given by

$$\omega_2(P) \equiv \omega_{\partial P} = \operatorname{sgn} \chi(g_{\partial P}) \tag{4.59}$$

from a $\mathbb{Z}_2$-valued 1-chain ω_1 $(\omega_1(\langle xy\rangle) \equiv \omega_{xy})$. ω_1 is determined up to a $\mathbb{Z}_2$ gauge transformation. Defining h_{xy} by

$$g_{xy} = \omega_{xy} h_{xy} \tag{4.60}$$

the Wilson action becomes

$$- S_{Y.M.W.} = \frac{1}{g_o^2} \sum_P \omega_{\partial P} \chi(h_{\partial P}) \quad . \tag{4.61}$$

We may compute expectation values by summing over all $\mathbb{Z}_2$ gauge fields ω_1 and all SU(2) gauge fields $\{h_{xy}\}$ subject to the constraint

$$\chi(h_{\partial P}) \geq 0 \tag{4.62}$$

for all plaquettes P .

Let us consider now

$$\overset{o}{W}(C) \equiv \omega_C = \prod_{\langle xy\rangle \in C} \omega_{xy} \tag{4.63}$$

$\overset{o}{W}(C)$ is gauge invariant and therefore determined by $\{g_{xy}\}$; it is a $\mathbb{Z}_2$ Wilson loop (actually $\overset{o}{W}(C) = \prod_{P \in S} \operatorname{sgn} \chi(g_{\partial P})$ where $\partial S = C$)

<u>Lemma 4.21</u> :

$\overset{o}{W}(C)$ has perimeter decay for small g_o^2 in the model modified by the constraint (4.58).

Proof :

$$\langle \overset{o}{W}(C) \rangle_{\Lambda} = \frac{1}{Z_{\Lambda}} \prod_{\langle xy \rangle} \int dh_{xy} \theta(\{h_{xy}\}) \sum_{\{\omega_{xy}\}} \omega_C \times$$

$$\times \exp \left[\frac{1}{g_o^2} \sum_P \omega_{\partial P} \chi(h_{\partial P}) \right] \tag{4.64}$$

where

$$\theta(\{h_{xy}\}) = \begin{cases} 1 & \text{if } \chi(g_{\partial P}) \geq 0 \quad , \quad \forall P \\ 0 & \text{otherwise} \end{cases}$$

If we define the probability measure $d\mu$ by

$$d\mu(\{h_{xy}\}) = \text{const.}\ \theta(\{h_{xy}\}) \tilde{Z}(\{h_{xy}\}) \prod_{\langle xy \rangle} dh_{xy} \tag{4.65}$$

where

$$\tilde{Z}(\{h_{xy}\}) = \sum_{\{\omega_{xy}\}} e^{\frac{1}{g_o^2} \sum_P \omega_{\partial P} \chi(h_{\partial P})} \tag{4.66}$$

we have

$$\langle \overset{o}{W}(C) \rangle_{\Lambda} = \int d\mu(\{h_{xy}\}) \langle \overset{o}{W}_c \rangle_{h,\Lambda} \tag{4.67}$$

where $\langle \overset{o}{W}_c \rangle_{h,\Lambda}$ is the expectation value of a $\mathbb{Z}_2$ Wilson loop with random couplings $\frac{1}{g_o^2} \chi(h_{\partial p}) \geq 0$.

By the concavity of the logarithm

$$\log \langle \overset{o}{W}(C) \rangle_{\Lambda} \geq \int d\mu(\{h_{xy}\}) \log \langle \overset{o}{W}_c \rangle_{h,\Lambda} \quad . \tag{4.68}$$

For the last expression the usual low temperature expansion can be carried out as in Theorem 3.14. The activities of defects are not necessarily small for all $\{h_{xy}\}$ but their μ-averages are, provided g_o^2 is small and convergence is established easily.

Remarks : (1) This does not prove that the usual Wilson loop $\langle W(C) \rangle = \langle \chi(h_C) \overset{o}{W}(C) \rangle$ has perimeter decay. It is quite plausible that larger defects will still sufficiently disorder the system to produce an area law for it.

(2) Mack and Petkova prove the dual statement

$$|\langle D_\omega(C)\rangle| \leq e^{-\alpha_\omega A(C)} \quad , \; \alpha_\omega > 0$$

by a similar method using the chessboard bound (Theorem 2.2) in an essential way.

Finally let us briefly explain 't Hooft's duality relation [76] for four-dimensional lattice Yang-Mills models. We discussed already vortex containers (or magnetic flux containers) and the importance of the free energy of magnetic flux. Instead of the containers of topology $S^2 \times S^2$ used above 't Hooft simply considers a system living in a 4-torus $S^1 \times S^1 \times S^1 \times S^1 \equiv T_4 \equiv \Lambda$. Magnetic flux is introduced by singular gauge transformations as follows : Pick two factors S^1_μ, S^1_ν from T_4, choose an element $\omega_{\mu\nu}$ in the center of G and a plaquette P_o in $S^1_\mu \times S^1_\nu$. Modify the action by multiplying $g_{\partial p}$ by $\omega_{\mu\nu}$ for each plaquette P in T_4 that is mapped onto P_o by the canonical projection from T_4 to $S^1_\mu \times S^1_\nu$. This may be done for all pairs (μ,ν) at the same time and the resulting modified partition function

$$Z_\Lambda(\{\omega_{\mu\nu}\})$$

can easily be seen to be independent of the choice of P_o. We say that $Z_\Lambda(\{\omega_{\mu\nu}\})$ carries magnetic flux $\omega_{\mu\nu}$ in the direction $*(\mu,\nu)$ dual to (μ,ν).

Electric flux is introduced in a dual way : Let $\{\chi_\tau\}_{\tau\in\hat{Z}}$ be a complete set of irreducible characters of the center Z of G. We define

$$\hat{Z}_\Lambda(\{\tau_{\mu\nu}\}) \equiv \int \chi_{\tau_{\mu\nu}}(\omega_{\mu\nu}) Z(\{\omega_{\mu\nu}\}) \prod_{\mu<\nu} d\omega_{\mu\nu} \tag{4.69}$$

and we say that $\hat{Z}_\Lambda(\{\tau_{\mu\nu}\})$ has electric flux $\tau_{\mu\nu}$ in the direction (μ,ν).

To motivate this language one should consider the response of (4.69) to another singular gauge transformation $\omega'_{\mu\nu}$: it gets multiplied by $\chi_{\tau_{\mu\nu}}(\omega'_{\mu\nu})$ and this is characteristic of electric flux (for instance a Wilson loop shows analogous behavior).

Now we specialize to $G = \mathbb{Z}_n$ or $SU(n)$ and look at the possible behavior of $Z(\{\omega_{\mu\nu}\})$ and $\hat{Z}(\{\tau_{\mu\nu}\})$ for increasing Λ. Denote the sides of Λ by $L_1,\ldots,L_4$.

Electric confinement corresponds to a bound

$$\frac{\hat{Z}_\Lambda(\tau_{12})}{\hat{Z}_\Lambda(\mathbb{1})} \leq c L_3 L_4 e^{-\alpha_{\tau_{12}} L_1 L_2} \tag{4.70}$$

where $\alpha_{\tau_{12}} > 0$ (the "string tension") and c are independent of $L_1,\ldots,L_4$, whereas magnetic confinement would mean for instance

$$\frac{Z_\Lambda(\omega_{34})}{Z_\Lambda(\mathbb{1})} \leq c' L_1 L_2 \, e^{-\beta_{\omega_{34}} L_3 L_4} \tag{4.71}$$

But the left hand side of (4.70) can be expressed in terms of the left hand side of (4.71) by (4.69)

$$\frac{\hat{Z}_\Lambda(\tau_{12})}{\hat{Z}_\Lambda(\mathbb{1})} = \frac{\sum_\omega e^{2\pi i \tau_{12}\omega} Z_\Lambda(\omega)/Z_\Lambda(\mathbb{1})}{\sum_\omega Z_\Lambda(\omega)/Z_\Lambda(\mathbb{1})} \tag{4.72}$$

and assuming (4.71) we get for $L_3, L_4 \to \infty$

$$\frac{\hat{Z}_\Lambda(\tau_{12})}{\hat{Z}_\Lambda(\mathbb{1})} \to 1$$

which is incompatible with (4.70). Note that (4.70) in fact implies a bound of the form

$$\left|\log \frac{Z_\Lambda(\omega_{12})}{Z_\Lambda(\mathbb{1})}\right| \leq c'' L_3 L_4 \, e^{-\alpha L_1 L_2} \tag{4.73}$$

where $\alpha = \inf_\tau \alpha_\tau$ and (4.73) in turn implies a bound of the form (4.70) (with $\alpha_{\tau_{12}} = \alpha$) . Note also that (4.73) exactly corresponds to (4.57) ; it expresses "spreading of magnetic flux".

We see again that spreading of magnetic flux is necessary for electric confinement. The analogy between this argument and the proof of Theorem 4.31 is visible.

The expansion results of Section 3 actually suggest a stronger version of the statement :

Conjecture : Whenever there is a mass gap and magnetic (electric) charges are not confined - i.e. the corresponding loop has perimeter decay - then electric (magnetic) charges are confined (cf. 't Hooft [76]).

It is noteworthy in this connection that the introduction of (electrically) charged matter fields destroyed the area law for the Wilson loop while producing an area law for the 't Hooft loop. This suggests again (by interchanging "electric" and "magnetic") that some kind of dynamical magnetic monopoles might be essential for electric confinement.

f) Some Rough Ideas About Roughening

Durhuus and Fröhlich not long ago published an article [66] containing such a wealth of ideas and results that we could not do justice to it by trying to present its contents in these lectures. They develop a very interesting picture based on considering a d-dimensional lattice Yang-Mills theory as a stack of (d-1)-dimensional nonlinear $G \times G$ σ-models with random couplings. The "spins" are the "vertical" gauge fields; their random coupling is provided by the "vertical" plaquettes. This allows to view the expectation of a Wilson loop as (imaginary) time evolution or diffusion of a string which thus traces out a random surface.

The most interesting aspect is this : For strong coupling the confinement bound follows simply from the uniform exponential clustering of the σ-models. But for $d \geq 4$ these σ-models have a critical point and a phase transition to an ordered state. The only way in which confinement can persist is then by wild fluctuations of the strings and the sheets they trace out because that leads to cancellations in the random phase factors. So a kind of "surface roughening" seems essential to maintain confinement.

Lately there has been a flurry of papers dealing with this presumed roughening transition (for instance [75]). Some of them give numerical evidence (breakdown of high temperature expansions), others offer intuitive evidence, mostly based on analogy with the Ising model. The most convincing argument, I think, has been proposed by Lüscher [75] in this context - it is well known in the context of the Ising model - and roughly goes as follows :

At high temperature (strong coupling) the cluster expansion for the Wilson loop gives a probability distribution of sheets (i.e. clusters of polymers in our language) spanned into the loop; there is a finite surface tension trying to keep the sheet from wiggling two much. Now at large coupling it should be reasonable only to consider simple sheets that may be described by a $\mathbb{Z}$-valued height function h in $d = 3$; in $d = 4$, h has to be $\mathbb{Z}^2$-valued. The effect of the surface tension might be subsumed in an effective action

$$K \sum_{\langle xy \rangle} (h_x - h_y)^2$$

where the sum is over nearest neighbors in $\mathbb{Z}^2$ (the plane of the Wilson loop). Now this is just the action of (the dual of) the Villain version of the plane rotator which for K small enough behaves essentially like a Gaussian - this is the famous Kosterlitz-Thouless transition [59] recently proven to exist [60]. But this means that the mean height fluctuation diverges logarithmically with the size of the loop: The surface gets completely delocalized! Actually it never really gets "rough" because the surface tension stays finite.

There can be no doubt that this effect really occurs, and it is a nuisance from the point of view of calculations : It probably would produce an essential singularity through which one has to pass when going from the well analyzed strong coupling region towards the critical point (which for continuous nonabelian group G and $d \leq 4$ presumably lies at $g_o = 0$). And the critical point is where the continuum has to be found ! So the sad fact is that there is no easy way to extrapolate the high temperature series to weak coupling. But in principle that should still be possible as long as there is no natural analyticity boundary in g_o which seems reasonable to believe.

So after these anticlimatic remarks let us turn to the continuum and the little that is known about it rigorously.

II. CONTINUUM GAUGE QUANTUM FIELD THEORIES.

This should be the main subject of these lectures, but unfortunately it is much less developed than lattice gauge theories. We will first describe the main approaches which fall into three classes : Continuum limits of lattice theories, direct continuum constructions and hybrid approaches. For the first two we will illustrate how they work in simple examples, namely pure Yang-Mills theory in two dimensions and the Schwinger model (massless QED_2). The hybrid appoach will concern us through much of the rest of these lectures : it is used in the construction of the two-dimensional abelian Higgs model that will be discussed in some detail and which will be shown to be a quantum field theory in the sense of Wightman; we also give an outline of the construction of massive QED_2 by this strategy. It will, however, become clear that Wightman's axioms are not the most natural framework for gauge theories, at least in the nonabelian case. So at the end we will discuss a possible alternative framework dealing with extended gauge invariant objects such as Wilson loops instead of local fields.

The philosophy behind this is the same that was already behind the lattice approximation : To avoid as much as possible to get entangled in the "hinterwelt" [1] of ghosts and unphysical degrees of freedom.

5. APPROACHES TO THE CONSTRUCTION OF CONTINUUM GAUGE QUANTUM FIELD THEORIES

a) The Scaling Limit.

This appears to be the most natural approach from a conceptual point of view; unfortunately it is hard to carry out in general because it requires detailed information about the behavior of lattice theories near a critical point. A detailed discussion for ϕ^4 models has been given by Schrader [2].

A lattice model is characterized by a number of "bare" parameters (coupling constants) $\underline{\lambda} = (\lambda_1,\ldots,\lambda_r) \in \mathbb{R}^r$ appearing in the action. Now one tries to find an equal number of "physical" parameters to characterize the theory, such as a number of masses expressed in physical units like kilograms or GeV's and maybe some dimensionless physical coupling constants (charges); we denote all these by

$$\underline{\mu} = (\mu_1,\ldots,\mu_r) \in \mathbb{R}^r \quad .$$

The model provides, for given lattice constant ε , a map, the "renormalization map" $R(\varepsilon) : \mathbb{R}^r \supset G \to \mathbb{R}^r$

$$R(\varepsilon)\underline{\lambda} \equiv \underline{\mu}(\varepsilon,\lambda) \qquad (5.1)$$

Now one wants to send ε to zero keeping $\underline{\mu}$ fixed; this means that all length scales measured in lattice units have to go to infinity, i.e. we have to move to a critical point. This requires of course some invertibility of the map $R(\varepsilon)$ which is the main concern of Schrader [2] in his study of ϕ^4 models.

Assuming this is possible and everything is sufficiently smooth, (5.1) implies a version of the Callan-Symanzik equations [3] :

$$0 = \frac{d}{d\varepsilon}\,\underline{\mu}(\varepsilon,\underline{\lambda}(\varepsilon)) = \sum_{i=1}^{r} A_i \frac{\partial\underline{\mu}}{\partial\lambda_i} + \frac{\partial}{\partial\varepsilon}\,\underline{\mu}(\varepsilon,\underline{\lambda}(\varepsilon))\Big|_{\lambda} \qquad (5.2)$$

where $A_i = \frac{\partial\lambda_i}{\partial\varepsilon}$ are usually again considered as functions of ε and $\underline{\mu}$.

The map $R(\varepsilon)$ is particularly simple for pure Yang-Mills theories because they depend only on one bare parameter g_o . It is convenient and customary to take as the corresponding physical quantity the "string tension"

$$\tilde{\alpha}_\tau \equiv - \lim_{A(C)\nearrow\infty} \frac{1}{\varepsilon^2 A(C)} \log\langle W_\tau(C)\rangle \qquad (5.3)$$

where $A(C)$ is the area measured in lattice units.

$\tilde{\alpha}_\tau$ depends on ε explicitly only in the obvious way

$$\tilde{\alpha}_\tau = \frac{1}{\varepsilon^2}\,\alpha_\tau \qquad (5.3')$$

where α_τ depends only on g_o , not on ε . (5.2) becomes

$$0 = -2\alpha_\tau - \beta(g_o)\,\frac{d\alpha_\tau}{dg_o} \qquad (5.4)$$

where β is the "lattice Callan-Symanzik function" $\beta = -\varepsilon\frac{dg_o}{d\varepsilon}$ considered as a function of g_o .

The hope is that the scaling limits of expectation values of reasonable functions of the fields

1) exist (this requires of course some action of the scale transformation on the fields).

2) are independent of the choice of physical parameters (within some reasonable class) to be held fixed.

3) are euclidean invariant and define a relativistic quantum field theory of some sort.

The third point could be established for instance by verifying the Osterwalder-Schrader axioms [4] for the expectation values of (local) gauge invariant fields or the assumptions on non-local gauge invariants objects formulated in Section 8. Warning : The second point is not always true as has been shown for the pure U(1) lattice gauge theory in three dimensions in the beautiful work of Göpfert and Mack [72].

Let us see how these ideas work in the trivial case of pure Yang-Mills theory in two dimensions. In order to get simple formulas (no Bessel functions) let us use the Villain form (1.8). By direct computation using the Peter-Weyl theorem [5] we obtain

$$\langle W_\tau(C)\rangle = e^{-\frac{1}{2} g_o^2 C_\tau A(C)} \tag{5.5}$$

(each link $\langle xy\rangle$ belongs at most to two plaquettes, so integration over g_{xy} requires the two adjacent plaquettes to carry the same irreducible representation σ as long as $\langle xy\rangle \notin C$. For free boundary conditions we obtain $\sigma = \mathbb{1}$ outside and $\sigma = \tau$ inside the loop; for other boundary conditions the thermodynamic limit also only lets these representations survive). So we have

$$\alpha_\tau = \frac{1}{2} g_o^2 C_\tau \tag{5.6}$$

$$\tilde{\alpha}_\tau = \frac{1}{2} \left(\frac{g_o}{\varepsilon}\right)^2 C_\tau \tag{5.7}$$

and we obtain a scaling limit by requiring

$$g_o(\varepsilon) = \varepsilon g \quad . \tag{5.8}$$

This corresponds to

$$\beta(g_o) = -g_o \quad . \tag{5.9}$$

(5.8) and (5.9) express the trivial "asymptotic freedom" of this superrenormalizable model.

It is easy to write down the continuum expectation value of any product of non-intersecting Wilson loops (the scale transformation acts on the loop observables simply by keeping their size fixed in physical units) :

$$<\prod_{i=1}^{N} W_{\tau_i}(C_i)> = \exp[-\frac{1}{2} g^2 \sum_{i=1}^{N} C_{\tau_i} A(C_i)] \tag{5.10}$$

and even overlapping Wilson loops cause no problem except requiring some Clebsch-Gordan gymnastics (see [6]). (5.10) is clearly euclidean invariant; of course there are no Osterwalder-Schrader axioms to be checked because we do not have expectation values of fields living at points, but rather on loops. In the last section I will discuss some modification of the Osterwalder-Schrader axioms appropriate for this situation.

Let us return to (5.4) for a moment. It is generally believed that the leading behavior (the first two nonvanishing Taylor coefficients) of the β-function is universal, i.e. independent of the way in which the scale or cutoff parameter is defined. Assuming this one can use the behavior of β as given by the famous computations of "asymptotic freedom" [7] in $d = 4$ that were done in the continuum by standard perturbation theory :

$$\beta(g_o) = -c_G g_o^3 + \mathcal{O}(g_o^5) \tag{5.11}$$

where c_G is proportional to the quadratic Casimir operator of G in the adjoint representation. Its precise value is determined by comparing the normalizations in (1.6) and (1.5) (see [8]).

This allows to predict the weak coupling behavior of the string tension when inserted in (5.4) :

$$\alpha_\tau = \mathcal{O}(g_o^q \exp - \frac{1}{c_G g_O^2})$$

It is truly remarkable that M. Creutz [8] in his celebrated Monte-Carlo studies of lattice Yang-Mills theory actually found at least consistency with this asymptotic behavior with the correct constant c_G . This is considered one of the strongest pieces of evidence for the soundness of the lattice approximation and for the correctness of the general belief that QCD confines quarks (a more sceptical view, however, seems to be suggested by [73]).

These discussions show how renormalization group ideas naturally enter into the study of the continuum limit. Bałaban [9] has announced a result that would go a long way towards controlling the continuum limit of 3-dimensional models and that makes use of the ideas of the renormalization group in an even more essential way, inspired by the stability proof for ϕ^4 model in 3 dimensions due to Gallavotti and friends [10]. The stability result announced by Bałaban has the form

$$|\log Z_\Lambda| \leq C|\Lambda|$$

with a constant C that is independent of the lattice spacing ε. His method is based on an iteration of a "block spin transformation" that produces a coarser and coarser lattice with increasingly strong coupling which is asymptotic freedom looked at from the opposite direction. One then starts this iteration at an increasingly fine lattice with smaller and smaller coupling in such a way that after a high enough number of iterations one essentially recovers the theory considered in the beginning. Since a detailed account is not yet available we will not discuss this approach beyond these remarks.

Note added : In the meantime a preprint [9] appeared in which Bałaban describes how his method works for a somewhat simpler model.

b) Direct Continuum Constructions.

It is sometimes possible to follow the more traditional methods of Constructive Quantum Field Theory also in the case of gauge models (see [11, 12, 69] for more information about Constructive QFT). For instance a version of QED_2 (the massive Schwinger-Thirring model) was constructed and established as a Wightman field theory already some time ago [13] by these methods. This was done however by a trick : The so-called bosonization maps this model into an ordinary scalar selfinteracting field theory (the massive Sine-Gordon model) and the standard constructive methods become applicable. The gauge aspect, however, is completely obscured by this trick, which furthermore has no chance of working in more than two dimensions.

A more generic approach can be followed for QED_2 (without the Thirring term). It is based on the Matthews-Salam formula [14] that proposes to construct the functional measure by first "integrating out" the fermion fields. This can be done directly in the continuum, provided some care is taken, and has been done successfully for Yukawa models [15, 16] (see [12] for more references). There is a little problem, however, because in the continuum one usually needs at first a cutoff (replacing the lattice cutoff) and these cutoffs have the annoying tendency to spoil either gauge invariance or Osterwalder-Schrader positivity. However, one may argue that it is sufficient to recover all desired properties in the limit when the cutoff is removed. Magnen and Sénéor have announced a partial construction of QED_3 based on this strategy [17].

The Matthews-Salam formulae can be obtained from the formalism described in Section 1 by integrating out the fermions following Berezin's prescription given there and passing to a formal continuum limit. A detailed discussion of this procedure is contained in [74]. It is convenient to rescale the gauge field $A \to eA$.

Gauge field expectations are then given by the formal probability measure

$$d\mu(A) = \frac{1}{Z} \det(1 + \frac{e}{i\not{p}+M} \not{A}) dm_{Y.M.}(A) \tag{5.13}$$

where $dm_{Y.M.}(A)$ is the (formal) continuum Yang-Mills measure. Expectations of fermion fields are also expressed by $d\mu$; a typical example is

$$\langle \bar{\psi}(x)\, e^{ie \int_x^y A dx'} \psi(y) \rangle =$$

$$= \int \operatorname{tr} e^{ie \int_x^y A dx'} (i\not{p}+e\not{A}+M)^{-1}(y,x) dm_{Y.M.}(A) \tag{5.14}$$

(trace over Dirac and internal indices).

In principle (5.13) and (5.14) are as good a heuristic starting point as the usual euclidean functional integral; one only has to verify the axioms in the end. Osterwalder-Schrader positivity is somewhat nonobvious, but it can be proven by a "Hamiltonian" derivation of (5.14) and (5.13) as given (with a correctible error, cf. [74]) in [18] for a Yukawa model, or it can be deduced by going through the lattice approximation in a more careful way; this will be done in the next section.

The message of (5.13) and (5.14) is the following : One should first study the integrands which correspond to external field problems and then worry about integrating over the gauge fields; in the abelian case we only have to integrate with a suitable Gaussian measure (that may contain cutoffs, gauge fixes etc.). First let us discuss the determinant in (5.13). It is of the form $\det(1+eK(A))$ where $K(A)$ can be taken to be

$$K(A) = (p^2+M^2)^{-3/4}(-i\not{p}+M)\, \not{A}(p^2+M^2)^{-1/4} \tag{5.15}$$

by a formal similarity transformation. We think of $K(A)$ as acting on $\mathcal{H}_F \equiv L^2(\mathbb{R}^d) \times \mathcal{V}_F$. For a sufficiently nice external field A_μ , $K(A)$ will be a compact operator in $\mathcal{I}_q$ for $q > q_o$, where

$$\mathcal{I}_q \equiv \{C | C \text{ compact operator on } \mathcal{H}_F \,;\, \operatorname{Tr}(C^*C)^{q/2} < \infty\}$$

<u>Lemma 5.1.</u>: Let $A_\mu \in \bigcap_{p\geq 1} L^p(\mathbb{R}^d)$. Then $K(A) \in \mathcal{I}_q$ for $q > d$.

<u>Proof</u> : In [19] it is proven that operators of the form $C = f(x)g(\nabla)$, i.e. multiplication in p-space followed by multiplication in x-space obey the bound

$$\|C\|_q \equiv \text{const}\ \|f\|_q\ \|g\|_q \tag{5.16}$$

where $q \geq 2$ and $\|C\|_q \equiv (\text{Tr}\ C^*C^{q/2})^{1/q}$. Now $K(A)$ may be written as a product of unitaries, $|A|^{1/2}(p^2 + M^2)^{-1/4}$ and its adjoint, respectively, from which the assertion follows. □

The point is that for operators of the form $1+K$, $K \in I_q$ $(q \geq 1)$ there is a well developed theory of modified Fredholm determinants (see [20,74] :

<u>Definition 5.2</u> : Let $C \in I_q$. Then

$$\det_p(1-C) \equiv \det\left[(1-C)e^{\sum_{k=1}^{p-1} \frac{1}{k}\text{Tr}\ C^k}\right]$$

(for $p > q$).

<u>Remark</u> : The right hand side is of the form $\det(1+B)$ with $B \in I_1$ (trace class) and therefore is well defined. Formally (or rigorously for $\|C\| < 1$ $\log \det_p(1-C) = -\sum_{k\geq p} \frac{1}{k}\text{Tr}\ C^k$). $\det_p$ has a number of good properties; among them

<u>Lemma 5.3</u> : $\det_p(1-zC)$ is an entire function of order p in z ; it has zeros exactly at the inverse eigenvalues of C.

<u>Proof</u> : See [20]. □

Returning to fermion gauge theories, we should restore the terms $\text{Tr}\ C^k$ $(k< p)$ that were deleted in $\det_p$ as well as possible in order to define a renormalized determinant. It is not legitimate to leave them out since this would not correspond to local counterterms. Note that $\text{Tr}\ K(A)^k$ corresponds to a one-loop Feynman graph with external gauge fields at the corners

$\text{Tr}\ K(A)^k$ ⟷ 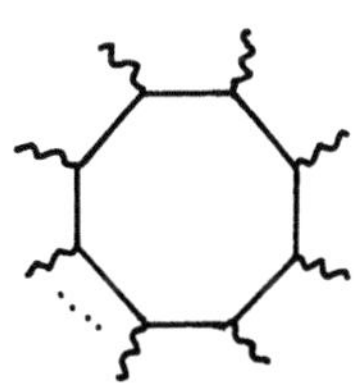

and we know very well how to renormalize these; denote the renormalized expressions by $\text{Tr}_{\text{ren}}K(A)^k$. Then we can define

<u>Definition 5.4</u> : $\det_{\text{ren}}(1+eK(A)) = \det_{d+1}(1+eK(A))\exp\left[-\sum_{k=1}^{d} \frac{e^k}{k}\text{Tr}_{\text{ren}}(-K(A))^k\right]$.

Remarks :

1) It is understood that the renormalization is done in such a way that $\det_{\mathrm{ren}}$ is gauge invariant and the subtractions correspond to local counterterms so that Osterwalder-Schrader positivity is preserved (all these things should be discussed in more detail; see [74]).

2) With a little work one can arrive at a closed expression for a renormalized determinant that is in fact based on a classic paper by Schwinger [21]; for instance in $d = 4$ it reads

$$\log|\det_{\mathrm{ren}}(1+eK(A))|^2 =$$

$$= \int_0^\infty dt \log(t\Lambda^2)\mathrm{Tr}(H_F(A)e^{-tH_F(A)} - e^{-tH_F(0)} H_F(0)) \tag{5.17}$$

Here Λ is a scale parameter (it has dimension of a mass) and $H_F(A)$ is a 4-dimensional nonrelativistic Pauli-Hamiltonian :

$$H_F(A) = (i\partial\!\!\!/+eA\!\!\!/+M)^*(i\partial\!\!\!/+eA\!\!\!/+M)$$

A detailed proof and discussion of (5.17) would lead too for afield here, but formally it is quite straightforward to derive (see [74] for a rigorous derivation).

Let us turn back to the Schwinger model. Its triviality stems from the fact that the determinant is so simple.

Lemma 5.5 : If $d = 2$, $A_\mu \in L^2$, $M = 0$ then

$$\det_4(1+eK(A)) = 1 \quad .$$

Corollary 5.6 : The spectrum of $K(A)$ is concentrated at zero, i.e. $K(A)$ is quasinilpotent.

Proof : This follows from the following three facts :

a) $\det_4(1+eK(A))$ is gauge invariant.

b) $\det_4(1+eK(A))$ is even in e .

c) $\det_4(1+eK(A)) = \det_4(1+ei\gamma_5 K(A))$.

a) is obvious by construction.

b) is the well known theorem of Furry; it is true because the determinant is invariant under charge conjugation which, however, changes e into $-e$.

c) follows from b) :

$$\det_4(1+eK(A)) = \det_2(1-eK(A)^2)^{1/2}$$

$$\text{and} \quad K(A)^2 = (i\gamma_5 K(A))^2 \; .$$

Now because of a) we may assume $\partial_\mu A_\mu = 0$ (replace A_μ by $A_\mu + \partial_\mu \Delta^{-1} \partial_\nu A_\nu$) . By c) and the fact $\gamma_5 \not{A} = i\not{B}$ with $B_\mu = \varepsilon_{\mu\nu} A_\nu$ we have $\det_4(1+eK(A)) = \det_4(1+eK(B))$. But B is a pure gauge, so by a) $\det_4(1+eK(B)) = \det_4(1) = 1$.

□

<u>Remark</u> : This is of course Schwinger's result [24]; this proof is taken from [23].

<u>Lemma 5.7</u> : For $d = 2$, $A_\mu \in L^2$, $M = 0$:

$$\det_{ren}(1+eK(A)) = \exp\left(-\frac{e^2}{2\pi} \|A^T\|_2^2\right)$$

where $A_\mu^T \equiv A_\mu + \partial_\mu \Delta^{-1} \partial_\nu A_\nu$ is the "transverse" part of A_μ .

<u>Proof</u> : This follows from the standard computation of $\mathrm{Tr}_{ren} K(A)^2$ (the second order vacuum polarization), cf. [24]; see also (7.64), (7.65).

□

<u>Remarks</u> :

1) $\det_{ren}(1+eK(A))$ obeys the "diamagnetic" bound of Section 1d) .

2) $\det_{ren}(1+eK(A))$ is <u>not</u> equal to $\prod_{i=1}^{\infty} (1+e\lambda_i)$ where $(\lambda_i)_{i=1}^{\infty}$ are the eigenvalues of $K(A)$. This is so even though no counterterm was required.

To complete the construction we have to give a meaning to the measure $dm_{Y.M.}(A)$. This will be a Gaussian measure with a covariance that has at first to contain a cutoff. A possibility is to take $dm_{Y.M.}^t$ with covariance $\hat{D}_{\mu\nu}(k) = \frac{1}{k^2}\left(\delta_{\mu\nu} - \frac{k_\mu k_\nu}{k^2}\right) e^{-tk^2}$. To make $dm(A)$ well defined, we also have to include a temporary space-time cutoff function in $K(A)$, for instance by replacing A_μ by $A_\mu g$ where $g \in C_0^\infty(\mathbb{R}^2)$. It is then straightforward to see that

$$\det_{ren}(1+eK(Ag)) \in L^1(dm_{Y.M.}^t)$$

and

$$d\mu^{t,g} \equiv \frac{1}{Z_{t,g}} \det_{ren}(1+eK(Ag))\, dm_{Y.M.}^t$$

is again a Gaussian measure with a covariance

$$\hat{D}^{t,g}_{\mu\nu}(k)$$

that converges in the weak sense (i.e. moments and characteristic function converge) towards $d\mu$ with covariance

$$\hat{D}_{\mu\nu} \equiv (k^2 + \frac{e^2}{\pi})^{-1}(\delta_{\mu\nu} - \frac{k_\mu k_\nu}{k^2}) \tag{5.18}$$

as $t \to 0$, $g \to 1$.

If we look at (5.18) we see that there is a free scalar field ϕ of mass $\frac{e}{\sqrt{\pi}}$ lurking in the background (as everybody knows of course). To see more precisely its rôle let us look at the current j_μ , determined by coupling in an external vector potential $a_\mu \in \mathcal{S}(\mathbb{R}^2)$ (Schwartz' test function space) with $\partial_\mu a_\mu = 0$:

$$
\begin{aligned}
&< e^{iej_\mu(a_\mu)} > \\
&= \lim_{\substack{t\to 0\\ g\to 1}} \frac{1}{Z_{t,g}} \int dm^t_{Y.M.}\ \det_{ren}(1+eK(gA+a)) \\
&= \int d\mu\, \exp(-\frac{e^2}{2\pi}\ \|a\|^2 - \frac{e^2}{\pi} A(a)) \\
&= \exp[-\frac{e^2}{2\pi}\ (f,(-\Delta+\frac{e^2}{\pi})^{-1}f)]
\end{aligned}
$$

where $f = \varepsilon_{\mu\nu}\partial_\mu a_u$. So j_μ behaves like a Gaussian random field defined by

$$j_\mu = \frac{1}{\sqrt{\pi}} \varepsilon_{\mu\nu}\partial_\nu\ (\phi+\theta) \tag{5.19}$$

with an arbitrary constant θ .

On the other hand $F = \varepsilon_{\mu\nu}\partial_\mu A_\nu$ is a Gaussian random field with covariance

$$\hat{C}(k) = \frac{k^2}{k^2 + \frac{e^2}{\pi}} = 1 - \frac{e^2}{\pi}\ (k^2 + \frac{e^2}{\pi})^{-1} \tag{5.20}$$

In a very symbolic way we can write therefore

$$F = \omega - i\frac{e}{\sqrt{\pi}}\ \phi \tag{5.21}$$

where ω is a white noise field (Gaussian generalized random field with covariance

$<\omega(f)\omega(g)> = \int f(x)g(x)dx^2)$. (5.21) is simply meant to express the relationship between the covariances of F , ω and ϕ . It cannot be taken literally, of course, because all three random fields are real valued.

A small surprise arises if we try to interpret F and j_μ jointly as random fields : This is not possible because the two-point function $<j_\mu F>$ is imaginary ! To see this we compute as before

$$<e^{iej_\mu(a_\mu)+iF(h)}>$$

$$= \int d\mu \exp[-\frac{e^2}{2\pi}\|a\|^2 - \frac{e^2}{\pi} A(a)+iF(h)]$$

$$= \exp\{-\frac{e^2}{2\pi}(f,(-\Delta+\frac{e^2}{\pi})^{-1}f)-\frac{1}{2}(h,(-\Delta)(-\Delta+\frac{e^2}{\pi})^{-1}h)$$

$$+ i\frac{e^2}{\pi}(f,(-\Delta+\frac{e^2}{\pi})^{-1}h)\} \tag{5.22}$$

So F , j_μ behave like Gaussian random fields as far as the "algebraic" structure is concerned, but the would-be covariances are

$$<j_\mu(x)j_\nu(x)> = \frac{1}{\pi}(\delta_{\mu\nu}+\Delta^{-1}\partial_\mu\partial_\nu)(-\Delta+\frac{e^2}{\pi})^{-1}(x,y) \tag{5.23a}$$

$$<j_\mu(x)F(y)> = -\frac{ie}{\pi}\varepsilon_{\mu\nu}\partial_\nu(-\Delta+\frac{e^2}{\pi})(x,y) \tag{5.23b}$$

$$<F(x)F(y)> = (-\Delta(-\Delta+\frac{e^2}{\pi})^{-1})(x,y) \tag{5.23c}$$

Note that all this information is already encoded in (5.19) and (5.21).

If we define the axial current

$$j_\mu^5 = -\varepsilon_{\mu\nu}j_\nu \tag{5.24}$$

we read off the consequence

$$\partial_\mu j_\mu^5 - \frac{ie}{\pi}F = 0 \tag{5.25}$$

which is the well-known axial anomaly relation (cf. [53]).

Another consequence is

$$\partial_\mu F - ie\varepsilon_{\mu\nu}j_\nu = \partial_\mu\omega \tag{5.26}$$

which should be regarded as the quantized version of the inhomogeneous Maxwell's equation. The right hand side is only a "contact term" that does not show up in the Wightman functions; the i becomes understandable if we remember that $F = iE$ where E is the physical electric field.

Since the equations (5.25) and (5.26) involve already i's and since analogous equations have to hold in any gauge theory with fermions (for instance QED_4 or QCD_4) we have to expect that such models can never be completely described in terms of random fields; formally they involve complex measures (I do not want to enter into a discussion of whether they can be genuine measures). In the language of Constructive Quantum Field Theory : Nelson-Symanzik positivity cannot be expected to hold in gauge theories with fermions.

θ-states can be defined by picking a value in (5.19) or, equivalently, by the prescription given in Section 4, namely by inserting a factor

$$e^{i\theta \int_\Lambda \varepsilon_{\mu\nu}\partial_\nu A_\mu} \equiv e^{i\theta \int_\Lambda F}$$

into the measure and sending $\Lambda \nearrow \mathbb{R}^2$.

Notice that there is no symmetry corresponding to $\phi \to \phi + \theta$ which would be axial U(1) symmetry. The conventional periodicity in θ comes from the fact that one usually considers periodic functions in ϕ to be naturally generated by the fields of the Schwinger model (such as : $\exp \frac{i}{\sqrt{4\pi}} \phi$: corresponding to : $\bar{\psi} e^{i\alpha\gamma_5}\psi$: in the bosonization scheme).

I do not want to extend this discussion of a trivial model too much (even though "the Schwinger model is inexhaustible" [25]); what I sketched here is only intended to convince the reader that most of the well known results may be recovered without entering any "Hinterwelt" [1].

The final strategy to be discussed is a mixed approach treating matter and gauge fields in a different way. Matter fields are at first put on a lattice and coupled to a continuum gauge field with some regularity properties corresponding to a cutoff on the Yang-Mills measure (only in the abelian case do we have a good way of doing this). Then we take the continuum limit for the matter fields pointwise in the gauge fields (i.e. the gauge fields are treated as external fields). Finally one has to remove the gauge field cutoff.

This procedure appears clumsy but it has proven so far the most flexible one. In particular it allows to prove Osterwalder-Schrader positivity and euclidean invariance, as well as the rest of the O.S. axioms in two dimensions for abelian models. The scheme has been applied by Challifour and Weingarten [26] and Weingarten [27] to the massive Schwinger model where it allowed to construct the theory at least in a finite volume, as well as to the two-dimensional Higgs (Landau-Ginzburg) model by Brydges, Fröhlich and myself [28,29]. The latter work will be the main topic of the following two sections and we will also remark how our methods work for QED_2. In Section 6 we treat external field problems whereas Section 7 is devoted to the removal of all cutoffs in the complete theory and verification of the axioms.

CONVERGENCE TO THE CONTINUUM LIMIT IN EXTERNAL OR CUTOFF GAUGE FIELDS.

External gauge fields coupled to a $P(\phi)_2$ model were studied by Schrader [30] and the analogue for ϕ_3^4 by Potthoff [31]. Their work, while fully satisfactory if one is only interested in the external field problem, is however not easily extendable to a fully quantized theory because

1) they do not include terms corresponding to virtual pair creation by the gauge field (i.e. the determinants to be studied below)

2) the selfinteraction of the scalar field which they use depends nonpolynomially on the gauge field.

In Ref. [26] a problem (QED_2) with an external gauge field that is "rough" is studied, i.e. the cutoff on the gauge field is removed from the beginning. This would be extremely hard in the case of bosonic matter and it is not really needed as will become clear. So here we follow Brydges, Fröhlich and myself [28] where Hölder continuous external gauge fields are used as a first step.

We show convergence for three types of objects : Covariant Green's functions, determinants and, finally, expectation values in a given external field. Finally we discuss quantized gauge fields with a cutoff.

a) Convergence of bosonic Green's functions [28].

To a large extent the continuum limit can be reduced to the limit of Feynman graphs. Therefore it is important to show convergence for the Euclidean propagators (Green's functions). The strategy followed here for Bose matter reduces this to termwise convergence of the perturbation series in the external gauge field; this is possible because we have a uniform bound on the Green's functions due to diamagnetism and sufficient analyticity in the coupling constant.

Given a lattice gauge field $\{g^{\varepsilon}_{xy}\}$ on $\varepsilon\,\mathbb{Z}^d$ the covariant Laplacean Δ^{ε}_g is defined as an operator on

$$\ell^2(\varepsilon\,\mathbb{Z}^d; V_H) \tag{6.1}$$

or, depending on the context,

$$\ell^2(\varepsilon\,\mathbb{Z}^d \cap \Lambda; V_H) \tag{6.2}$$

where Λ is a bounded open set in $\mathbb{R}^d$ and $\varepsilon\,\mathbb{Z}^d$ is considered imbedded in $\mathbb{R}^d$. (6.1) and (6.2) consist of square summable functions from the lattice into the Higgs vector space V_H. Δ^{ε}_g is determined by the quadratic form

$$-(\phi,\Delta^{\varepsilon}_g\phi) = \sum_{<xy>} \|\phi(x)-U_H(g_{xy})\phi(y)\|^2 \tag{6.3}$$

(where ϕ is a Higgs field of compact support) and taking the Friedrichs extension (cf. [33]).

The Green's function is the kernel of $(-\Delta^{\varepsilon}_g + m^2)^{-1}$, i.e.

$$C^{\varepsilon}_g(x,y) \equiv (-\Delta^{\varepsilon}_g + m^2)^{-1}(x,y) \tag{6.4}$$

which takes values in $L(V_H)$, the linear operators on V_H.

First we note the following simple diamagnetic bound ([32]; Refs [27] through [30] of Part I) :

Theorem 6.1 : Let $\varphi \in V_H$. Then

$$|(\varphi,C^{\varepsilon}_g(x,y)\varphi)| \leq (\varphi,C^{\varepsilon}_{\mathbb{1}}(x,y)\varphi)$$

Proof : This is almost trivial. By (6.1)

$$-\Delta^{\varepsilon}_g + m^2 = 2d + m^2 - A_g$$

where A_g "couples nearest neighbors" :

$$A_g(x,y) = \begin{cases} U_H(g_{xy}) & \text{if } x,y \text{ are nearest neighbors} \\ 0 & \text{otherwise} \end{cases}$$

Obviously $\|A_g\| \leq 2d$. So

$$(-\Delta_g^\varepsilon+m^2)^{-1} = (2d+m^2)^{-1}(1 - \frac{A_g}{2d+m^2})^{-1} =$$

$$= (2d+m^2)^{-1} \sum_{n=o}^{\infty} (\frac{A_g}{2d+m^2})^n$$

where the series converges in operator norm.

Now $A_g^n(x,y)$ is a sum over paths of length εn from x to y of products of unitaries (the parallel transporters from x to y). Hence

$$|(\varphi,A_g^n(x,y)\varphi)| \leq (\varphi,A_{\mathbb{1}}^n(x,y)\varphi)$$

from which the theorem follows.

<u>Remark</u> : The analogous statement is false for fermions (due to their paramagnetism): Assume it were true. Then, using the notation explained after eq. (6.24),

$$\log \det \frac{\not{D}_{A^\varepsilon}^\varepsilon+R_{A^\varepsilon}+m}{\not{\partial}^\varepsilon+R_o+m} = \operatorname{Tr} \log \frac{\not{D}_{A^\varepsilon}^\varepsilon+R_{A^\varepsilon}+m}{\not{\partial}^\varepsilon+R_o+m} =$$

$$= -\lim_{T\to\infty} \int_o^T dt \operatorname{Tr}[(\not{D}_{A^\varepsilon}^\varepsilon+R_{A^\varepsilon}+m+t)^{-1} - (\not{\partial}^\varepsilon+R_o+m+t)^{-1}]$$

would be ≥ 0 . But by Theorem 2.3 it is ≤ 0 which is only compatible if it vanishes which of course does not happen in general. This remark shows that the "diamagnetic" inequality (Theorem 2.3) for fermions is really an expression of paramagnetism.

So for convergence of fermion Green's functions one has to find a different method. But because the usual fermion models do not involve direct self interaction of the fermions one can use the Matthews-Salam formalism (see above) which only involves products of Green's functions with determinants. These products can be bounded easily (see [15]); we will come back to this in the subsection about determinants.

Next we turn to the analyticity result alluded to before. It is convenient to consider a slightly more general class of Green's functions by replacing $U_H(g_{xy})$ with arbitrary bounded functions $1+A$ from the links into $L(V_H)$. The corresponding Green's functions are denoted by

$$C_A^\varepsilon(x,y) \tag{6.5}$$

and we use the norm

$$\|A\| \equiv \sup_{\langle xy \rangle} \|A_{xy}\| \quad . \tag{6.6}$$

Then we have the following result :

<u>Lemma 6.2</u> : Let A,B be bounded $L(V_H)$ valued functions of the links of $\Lambda \cap \varepsilon \mathbb{Z}^d$ and χ_Λ the characteristic function of Λ . Then $\chi_\Lambda(x)C_{A+\lambda B}(x,y)\chi_\Lambda(y)$ is a real analytic function of λ with values in $\ell^2((\Lambda \cap \varepsilon \mathbb{Z}^d) \times (\Lambda \cap \varepsilon \mathbb{Z}^d) \; ; \; L(V_H))$. It extends to a function analytic in the strip defined by

$$\frac{2}{m}|\mathrm{Im}\,\lambda| \; \|B\| + \left(\frac{\mathrm{Im}\,\lambda}{m}\right)^2 \|B\|^2 \equiv \xi < 1$$

and its extension $\chi_\Lambda \tilde{C}_{A+\lambda B}\chi_\Lambda$ obeys

$$\| \chi_\Lambda \tilde{C}_{A+\lambda B}\chi_\Lambda \|_2 \leq \| \chi_\Lambda C_{A+B\,\mathrm{Re}\,\lambda}\chi_\Lambda \|_2 \frac{1}{1-\xi} \tag{6.7}$$

<u>Remark</u> :

1) C itself is not analytic because of the norms appearing in (6.3).

2) The norms in (6.7) are Hilbert-Schmidt norms = ℓ^2-norms of the kernels.

<u>Proof</u> : The lemma is proven by expanding $\chi_\Lambda C_{A+\lambda B}\chi_\Lambda$ in a power series in $\mathrm{Im}\,\lambda$; convergence is obtained in the strip described above by essentially routine methods. See [28] for more details.

□

The relevance of this lemma becomes clearer through

<u>Corollary 6.3</u> : Let A^ε be a function of the links of $\varepsilon \mathbb{Z}^d \cap \Lambda$ taking values in the Lie algebra of G . By the exponential map this induces a gauge field $\{g^\varepsilon_{xy}\}$: $g^\varepsilon_{xy} \equiv \exp i\varepsilon A^\varepsilon_{xy}$ (Here we use the physicists' convention of antihermitian Lie algebra elements). Define $g^\varepsilon_{xy}(\lambda) \equiv \exp i\varepsilon\lambda A^\varepsilon_{xy} (\lambda \in \mathbb{R})$. Then $\chi_\Lambda C_{g(\lambda)}\chi_\Lambda$ is real analytic in λ and extends to a function $\tilde{C}^{\varepsilon,\Lambda}_{g(\lambda)}$ analytic in a domain defined by

$$\frac{2|\mathrm{Im}\,\lambda|}{m} f(\varepsilon,\lambda,\|A^\varepsilon\|) + \frac{(\mathrm{Im}\,\lambda)}{m^2} f(\varepsilon,\lambda,\|A^\varepsilon\|)^2 \equiv \xi < 1 \tag{6.8}$$

where

$$f(\varepsilon,\lambda,\|A^\varepsilon\|) = \frac{1}{\varepsilon|\lambda|} e^{\frac{\varepsilon}{2}|\mathrm{Im}\,\lambda| \, \|A^\varepsilon\|} \times$$

$$\times \left[4\left(\mathrm{sh}\left(\frac{\varepsilon}{2}|\mathrm{Im}\,\lambda| \, \|A^\varepsilon\|\right)\right)^2 + \varepsilon^2(\mathrm{Re}\,\lambda)^2 \|A^\varepsilon\|^2\right]^{1/2}$$

and obeys the bound

$$\|\tilde{C}^{\varepsilon,\Lambda}_{g(\lambda)}\|_2 \leq \|\chi_\Lambda C^\varepsilon_1 \chi_\Lambda\|_2 (1-\xi)^{-1} \quad . \tag{6.10}$$

Remark : Note that the complicated looking region described in (6.8), (6.9) for $\varepsilon \to 0$ reduces to the strip

$$\frac{2|\mathrm{Im}\ \lambda|}{m} \|A\| + \frac{(\mathrm{Im}\ \lambda)^2}{m^2} \|A\|^2 \equiv \xi < 1 \ , \tag{6.11}$$

provided $\lim_{\varepsilon\to 0}\|A^\varepsilon - A\| = 0$ and it contains a strip of width independent of ε .

Proof : We use the lemma with

$$A_{xy} \equiv \frac{1}{\varepsilon}(e^{i\varepsilon A^\varepsilon_{xy}} - \mathbf{1})$$

$$B_{xy} \equiv \frac{1}{\varepsilon\lambda} e^{i\varepsilon A^\varepsilon_{xy}}(e^{i\varepsilon\lambda A^\varepsilon_{xy}} - \mathbf{1}) \quad .$$

The fact that B itself depends on λ does not cause any problems (the composition of holomorphic functions is holomorphic). To obtain (6.10) we used in addition diamagnetism in the form of Theorem 6.1. □

Before we state the main result of this subsection we have to make a trivial but crucial remark and a definition.

Remark : There is a natural imbedding Q^*_ε of lattice ℓ^2 into continuum L^2 by identifying lattice functions with piecewise constant functions. The adjoint Q_ε maps L^2 into ℓ^2 by averaging functions over elementary (hyper-) cubes. Q^*_ε is an isometry, Q_ε a partial isometry. When we speak about convergence of lattice functions to continuum functions we always tacitly assume the use of this embedding.

Definition 6.4 : A family $\{g^\varepsilon_{xy}\}$ of lattice gauge fields is said to converge to a continuum gauge field A_μ as $\varepsilon \to 0$ iff

$$A^\varepsilon_\mu(x) \equiv \frac{1}{i\varepsilon}(U_H(g^\varepsilon_{x,x+\hat{e}_\mu}) - \mathbf{1}) \tag{6.12}$$

converges to A in L^∞ norm ($\hat{e}_\mu$ is the unit vector in $+\mu$ direction). A typical example of this convergence occurs when the lattice gauge fields are obtained by integrating a continuum gauge field. We have the following result :

Theorem 6.5 : Assume that the lattice fields $\{g_{xy}\}$ converge to A_μ as $\varepsilon \to 0$. Then the Green's functions $\chi_\Lambda C^\varepsilon_{g^\varepsilon} \chi_\Lambda$ converge to $\chi_\Lambda C_A \chi_\Lambda$ in $L^p(\Lambda \times \Lambda)$ for $1 \leq p < \frac{d}{d-2}$ in $d = 2,3$. C_A denotes the continuum Green's function $(-\Delta_A + m^2)^{-1}(x,y)$.

Remark : This result probably also holds for $d > 3$. A proof would require a generalization of Lemma 6.2 with the ℓ^2 topology replaced by ℓ^p topology $(p < \frac{d}{d-2})$. This has not yet been done.

Proof :

1) L^2 convergence implies L^p convergence trivially for $p \leq 2$; for $p \geq 2$ it is also true because of Hölder's inequality and the fact that $\|C^\varepsilon_g\|_{L^p}$ is uniformly bounded for $\varepsilon \to 0$ by the diamagnetic bound, provided $p < \frac{d}{d-2}$.

2) Corollary 6.3 tells us that we may assume the gauge field $A^\varepsilon_\mu(x)$ to be arbitrarily small, using an argument parallel to the proof of Vitali's theorem :

Given $\delta > 0$, we can find a λ_o that brings $g(\lambda_o)$ close enough to $\mathbb{1}$ to make $\|A^\varepsilon_\mu\| < \delta$ for all $\varepsilon < 1$ (see Corollary 6.3 and (6.12). Suppose we know convergence as $\varepsilon \to 0$ for these "small" gauge fields. By Corollary 6.3 $\chi_\Lambda C^\varepsilon_{g^\varepsilon(\lambda)} \chi_\Lambda$ is analytic in a strip-like region and by (6.10) uniformly bounded. To see convergence at $\lambda = 1$ take a sequence of open disks $D_1, \ldots, D_N$, and contained in the strip, such that the center of D_{i+1} is in D_i $(i = 1, \ldots, N-1)$ and $D_1 \ni \lambda_o$, $D_N \ni 1$. Assuming convergence in $D_1, \ldots, D_k$ we get convergence of the Taylor coefficients of the expansion around the center of D_{k+1} and this gives convergence in D_{k+1}.

3) We choose now $\|A^\varepsilon_\mu\|$ so small that the perturbation expansion of $\chi_\Lambda C^\varepsilon_{g^\varepsilon} \chi_\Lambda$ in powers of A^ε_μ converges in ℓ^2 (uniformly in ε). It is easy to see that this is possible and that the terms of the series converge to the terms of the perturbation series for the continuum Green's function. For more details consult [28].

□

There is another nasty little thing whose convergence as $\varepsilon \to 0$ is needed, the so-called re-Wick ordering constant. It is well known that in the continuum the potential $V(|\phi|)$ has to be Wick ordered. This Wick ordering formally corresponds to subtracting terms with divergent coefficients such as $C(0)$. If an external gauge field is present it is much more convenient and certainly legitimate to use the covariance (= Green's function) C_A for Wick ordering. This is what Schrader [30] and Potthoff [31] did. If we think about "quantizing" the gauge field by integrating over it this becomes contrary to the spirit of renormalization theory; in any case it would correspond to a nonlocal nonpolynomial Lagrangean and there is little chance to verify Osterwalder's and Schrader's axioms. Therefore one has to discuss the re-Wick ordering terms

$$\delta C^{\varepsilon}_{g^{\varepsilon}}(x,x) \equiv C^{\varepsilon}_{g^{\varepsilon}}(x,x) - C^{\varepsilon}_{\mathbb{1}}(x,x) \tag{6.13}$$

ans show that they have finite limits as $\varepsilon \to 0$. This is true in $d = 2,3$:

Theorem 6.6 : If the family of gauge fields g^{ε} converges as $\varepsilon \to 0$, then

$$\int_{\Lambda} |\delta C^{\varepsilon}_{g^{\varepsilon}}(x,x) - \delta C^{\varepsilon'}_{g^{\varepsilon'}}(x,x)|^{p} dx \to 0$$

for $\varepsilon,\varepsilon' \to 0$ and $1 \leq p \leq \infty$ $(d = 2,3)$.

Remark : For $d = 4$ the re-Wick ordering is infinite and some extra renormalization is required (this is no reason for concern here because nobody knows how to construct a field theory in $d = 4$ anyhow). Typical divergent graphs that spoil Theorem 6.6 are

and they contribute to charge renormalization.

Proof : This is quite complicated and we refer to [28] where the case d = 2 is treated in detail. Here we only try to give the strategy.

To make the cancellations in $\delta C^{\varepsilon}_{g^{\varepsilon}}$ visible we write again $g^{\varepsilon}_{xy}(\lambda) = \exp i\varepsilon\lambda A^{\varepsilon}_{xy}$ and consider

$$F_x(\lambda) \equiv \delta C^{\varepsilon}_{g^{\varepsilon}(\lambda)}(x,x) \quad . \tag{6.13}$$

Note that

1) $\qquad F_x(0) = 0$

2) $\qquad F'_x(0) = 0 \quad .$ (6.14)

1) is obvious;

2) follows from the fact that F_x is even;

this can best be seen from an expansion in paths as used in the proof of Theorem 6.1.

Here we will have a sum over closed paths beginning and ending at x ; to each path there is also its inverse that corresponds to the opposite orientation; taking the inverse path has the same effect as replacing λ by $-\lambda$ (this is essentially again Furry's theorem).

Because of (6.14) we have

$$F_x(1) = \int_0^1 d\lambda(1-\lambda)F''_x(\lambda) \quad . \tag{6.15}$$

To see what this is, write again

$$A^{\varepsilon}_{xy}(\lambda) = \frac{1}{i\varepsilon}(U_H(g^{\varepsilon}_{xy}(\lambda)) - \mathbb{1}) \tag{6.16}$$

$$\Delta^{\varepsilon}_{g^{\varepsilon}(\lambda)} \equiv \Delta^{\varepsilon}_{\mathbb{1}} + W^{\varepsilon} \tag{6.17}$$

which can be written symbolically as

$$W = i(d^*A + A^*d) - A^*A \tag{6.18}$$

(d is the lattice gradient, $d^* = \delta$ in the chain notation of Section 4: A can be considered as a map from 0-chains to 1-chains).

Now

$$F''_x(\lambda) = \frac{d^2}{d\lambda^2} C^{\varepsilon}_{g^{\varepsilon}(\lambda)}(x,x) \tag{6.19}$$

and

$$\frac{d^2}{d\lambda^2} C^{\varepsilon}_{g^{\varepsilon}(\lambda)} = 2\, C^{\varepsilon}_{\mathbb{1}} \frac{dW^{\varepsilon}}{d\lambda} C^{\varepsilon}_{\mathbb{1}} \frac{dW^{\varepsilon}}{d\lambda} C^{\varepsilon}_{\mathbb{1}} - C^{\varepsilon}_{\mathbb{1}} \frac{d^2W^{\varepsilon}}{d\lambda^2} C^{\varepsilon}_{\mathbb{1}} \ . \tag{6.20}$$

Inspection shows that (6.20) does not contain any divergent parts. A little functional analysis together with diamagnetism (Theorem 6.1) shows that $\delta C^{\varepsilon}_{g^{\varepsilon}(\lambda)} = F_x(1)$ has a uniformly (in ε) bounded L^p norm and is a L^1-Cauchy sequence for $\varepsilon \to 0$. This is enough to prove the theorem by Hölder's inequality. □

b) Convergence of determinants.

The importance of the convergence of determinants became clear already in our discussions of the Matthews-Salam formalism in Section 5. It is also important for the construction of Higgs models : Formally the measure

$$e^{-\frac{1}{2}\int|D\phi|^2-\frac{1}{2}m^2\int|\phi|^2}\prod_x d\phi(x)$$

is identical to the normalized Gaussian measure $d\mu(\phi)$ with covariance $C_A = (-\Delta_A+m^2)^{-1}$, multiplied by a factor that is a multiple of

$$\det\left(\frac{-\Delta_A+m^2}{-\Delta+m^2}\right)^{-1}$$

(On the lattice of course all these statements become true). Physically this factor contains the virtual pair creation by the gauge field.

So in this subsection we want to study both boson and fermion determinants. We will make use of the formalism of modified determinants and I_p spaces introduced in Section 5.

We always write the operators of which the determinants are to be taken in the form $1+K(A)$.

So for bosons we define

$$K^\varepsilon_H(A^\varepsilon) \equiv (C^\varepsilon_{\mathbb{1}})^{1/2}\, W^\varepsilon (C^\varepsilon_{\mathbb{1}})^{1/2} \tag{6.21}$$

with

$$W^\varepsilon = \Delta^c_{g^\varepsilon} - \Delta^c_{\mathbb{1}} \tag{6.22}$$

and

$$A^\varepsilon = \frac{1}{i\varepsilon}(U_H(g^\varepsilon) - \mathbb{1}) \tag{6.23}$$

For fermions we define in analogy with Section 5

$$K^\varepsilon_F(A^\varepsilon) \equiv (C^\varepsilon_{\mathbb{1},F})^{3/4}(\not{D}^\varepsilon_o+R_o+M)(\not{D}^\varepsilon_{A^\varepsilon}-\not{D}^\varepsilon_o+R_{A^\varepsilon}-R_o)+(C^\varepsilon_{\mathbb{1},F})^{1/4} \tag{6.24}$$

Here $\not{D}^\varepsilon_{A^\varepsilon}+R_{A^\varepsilon}$ is the lattice Dirac operator determined by

$$M\,\Sigma\bar{\psi}\psi+S_F = -K\,\Sigma\psi(\not{D}^\varepsilon_{A^\varepsilon}+R_{A^\varepsilon})\psi$$

and

$$C^\varepsilon_{\mathbb{1},F} = [(\not{D}^\varepsilon_o+R_{A^\varepsilon}+M)^*(\not{D}^\varepsilon_o+R_o+M)]^{-1}$$

so according to Section 1, if we write $e^{i\varepsilon P_\mu}$ for translation by 1 lattice unit in the $+\mu$-direction, g^ε_μ for $g_{x,x+\varepsilon\hat{e}_\mu}$, we have for $\theta = 0$

$$\not{D}^\varepsilon_{A^\varepsilon} + R_{A^\varepsilon} = \sum_\mu \frac{r+\gamma_\mu}{2\varepsilon} U_F(g^\varepsilon_\mu) e^{i\varepsilon P_\mu} + \sum_\mu \frac{r-\gamma_\mu}{2\varepsilon} e^{-i\varepsilon P_\mu} U^*_F(g^\varepsilon_\mu) - \frac{rd}{\varepsilon} \tag{6.25}$$

which becomes with the definition

$$A^\varepsilon = \frac{1}{i\varepsilon}(U_F(g^\varepsilon) - \mathbb{1}) \tag{6.26}$$

$$\begin{aligned} \not{D}^\varepsilon_{A^\varepsilon} + R_{A^\varepsilon} &= \frac{i}{\varepsilon}\gamma^\mu \sin\varepsilon P_\mu - \frac{i}{2}\gamma^\mu\{A^\varepsilon_\mu \cos\varepsilon P_\mu + \cos\varepsilon P_\mu (A^\varepsilon_\mu)^*\} - \frac{1}{2}\gamma^\mu\{A^\varepsilon_\mu \sin\varepsilon P_\mu - \sin\varepsilon P_\mu (A^\varepsilon_\mu)^*\} \\ &+ \frac{r}{\varepsilon}\sum_\mu(\cos\varepsilon P_\mu - 1) - \frac{r}{2}\{A^\varepsilon_\mu \sin\varepsilon P_\mu + \sin\varepsilon P_\mu (A^\varepsilon_\mu)^*\} + \frac{ir}{2}\{A^\varepsilon_\mu \cos\varepsilon P_\mu - \cos\varepsilon P_\mu (A^\varepsilon_\mu)^*\} \end{aligned} \tag{6.27}$$

For $\theta \neq 0$ we simply have to multiply R_{A^ε} and R_o by $e^{i\theta\gamma_5}$.

We can now state the following convergence result :

<u>Theorem 6.7</u> : $K^\varepsilon_H(A^\varepsilon)$ and $K^\varepsilon_F(A^\varepsilon)$ converge in I_p for any $p > d$ as $\varepsilon \to 0$, provided A^ε converges to A in L^∞ and A is a continuum Yang-Mills field of compact support.

<u>Remarks</u> :

1) All operators are assumed to act on $H_i \equiv L^2 \times V_i$, $i = H,F$.

2) The compact support requirement can be replaced by sufficiently strong (power-like) fall-off. Theorem and proof are essentially taken from [28], where however only the Higgs case in $d = 2$ is discussed. See [26] for a related "fermionic" result.

<u>Proof</u> : It is very easy to see that $K^\varepsilon_{H,F}(A^\varepsilon)$ are uniformly in I_p $(p > d)$ by writing them as a product of operators uniformy in I_{2p} ; again one uses the theorem on operators of the form $A = f(p)g(x)$ (multiplication in x-space by g followed by multiplication in p-space by f) used in Section 5 (see [19]) :

$$\|A\|_q \leq \text{const} \|f\|_q \|g\|_q \quad , \quad q \geq 2 \; .$$

To get I_p convergence it is sufficient to show I_{2p} convergence of $\chi_\Lambda(C^\varepsilon_{1,F})^{1/2}$ which is elementary, together with the following lemma :

<u>Lemma 6.8</u> : Let $\|A_n - A\|_p \to 0$; B_n a uniformly bounded sequence of operators such that $B_n \to B$, $B^*_n \to B$ strongly. Then $A_n B_n \to AB$ in I_p. For a proof of this lemma

see [28].

These remarks should make sufficiently clear how Theorem 6.7 is proven.

We are mostly interested in

Corollary 6.9 : Let A^{ε} converge to A in L^{∞} where A has compact support. Then $\det_p(1+K_i^{\varepsilon}(A^{\varepsilon}))$ converges as $\varepsilon \to 0$ for $p > d$ and $i = H$ or F.

Proof : This follows from the standard result about $\det_p$ asserting that $\det_p$ is Lipschitz continuous on I_p (cf. [20], [34]).

So all that is missing is convergence of the low order expressions $\mathrm{Tr}\, K_i^{\varepsilon}(A^{\varepsilon})^q$, $q \leq d$. The best we have to offer for this is explicit computation and renormalization where necessary. This is notoriously tedious - lattice Feynman graphs are no fun. It has been carried out in all detail for the second order vacuum polarization in QED_2 [26] and $Higgs_2$ [28]. The results in these cases, which do not require renormalization, agree with those obtained by other gauge invariant regularizations such as Pauli-Villars or dimensional regularization. The necessary information for $QED_{3,4}$ is contained in the careful study of Sharatchandra [77].

To rest safely, one should carry out the calculations or give some general argument for the nonabelian models in dimension $d > 2$. However, there is no fundamental obstacle preventing to make the following a theorem :

Quasi-Theorem 6.10 : Let $d = 4$. There are functions $g_i(\varepsilon)$ $(i = H,F)$ with $g_i(\varepsilon) \to 0$ as $\varepsilon \to 0$ such that

$$\det(1+K_i^{\varepsilon}(A^{\varepsilon})) \times \exp[-\frac{1}{g_i(\varepsilon)^2} \sum_p (\chi(g^{\varepsilon}_{\partial p})-\chi(\mathbf{1})]$$

converges as $\varepsilon \to 0$ to a nonzero limit

$$\det_{ren}(1+K_i(A))$$

(which is actually given in Definition 5.4 for fermions), provided $A^{\varepsilon} \to A$ in some sufficiently strong Sobolev topology. Even more straightforward should be

Quasi-Theorem 6.10' : In $d = 3$

$$\lim_{\varepsilon \to 0} \det(1+K_i^{\varepsilon}(A^{\varepsilon})) \equiv \det_{ren}(1+K_i(A))$$

exists for $i = H,F$ and is nonzero, provided $A^{\varepsilon} \to A$ in some sufficiently strong Sobolev topology. Finally there is a proven fact :

Theorem 6.11 : Let $d = 2$. Then $\lim_{\varepsilon\to o} \det(1+K_i^{\varepsilon}(A^{\varepsilon}))$ exists and is nonzero for $i = H,F$, provided $A^{\varepsilon} \to A$ in the norm

$$\|A\|_{\infty,\alpha} \equiv \|A\| + \int_{\Lambda\times\Lambda} dx\, dy \frac{\|A(x)-A(y)\|^2}{|x-y|^{2+\alpha}}$$

for some $\alpha > 0$. Λ is the support of A^{ε} , A .

Remark : The stronger topology as compared to Theorem 6.7 is only necessary to make the second order vacuum polarization converge. For $i = H$ this is proven in [28], for $i = F$ a slightly stronger result occurs in [26] (it is not, strictly speaking, stronger since it deals with a random field A and proves convergence only in some mean sense; A is, however, allowed to be much rougher, in fact it does not even have to be a function but may be a distribution).

Proof : We gave most of the arguments already. The norm used looks more plausible in Fourier space : It has the form

$$\|A\|_{\infty,\alpha} = \|A\| + C_{\alpha} \int |\hat{A}(k)|^2 |k|^{\alpha} d^2k$$

and is of course tailored to make $\int d^2k\, \hat{A}_{\mu}^{\varepsilon}(k)\hat{A}_{\mu}^{\varepsilon}(k) \times \Pi_{\mu\nu}^{\varepsilon}(k)$ converge where $\Pi_{\mu\nu}^{\varepsilon}$ is the second order vacuum polarization corresponding to the Feynman graphs

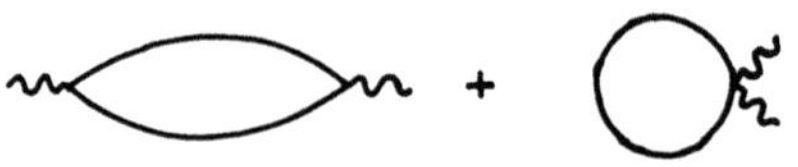

for $i = H$

for $i = F$

Explicit computation shows that $\Pi_{\mu\nu}(k) = O(\log|k|)$ for large k if $i = H$ and $O(1)$ for $i = F$.

□

This should also give sufficient ideas which Sobolev topology has to be chosen in $d > 2$.

The limit $\det_{ren}$ does not vanish because $\det_3(1+K_i(A))$ is a function that vanishes only if $K_i(A)$ has an eigenvalue -1 . This would correspond to a zero eigenvalue of $\not{D}_A + M$ (or $-\Delta_A + M^2$, respectively) which is impossible because

$$(\not{D}_A+M)^*(\not{D}_A+m) \geq M^2 > 0$$

(see [15] for a more rigorous but less transparent version of this argument). In [28] a more refined lower bound is derived.

Theorem 6.12 : In $d = 2,3$

$$\left|\det_{ren}(1+K_i(A))^{\varepsilon_i}\right| \leq 1$$

for $i = H,F$, $\varepsilon_H = -1$, $\varepsilon_F = 1$.

Proof : This follows directly from the "diamagnetic" inequality Theorem 2.3 by a limiting argument. See also [22].

□

Remarks :

1) In $d = 4$ the analogous statement is definitely false due to the counterterm that is needed. This must have been known already to Heisenberg and Euler in 1936 [68] in some form. See [74] for a detailed discussion.

2) By construction the renormalized determinants are gauge invariant, so they obey Ward identities such as

$$\partial_\mu \frac{\partial}{\delta A_\mu} \det_{ren}(1+K_i(A)) = 0$$

for a $U(1)$ theory or more generally

$$D_\mu(A) \frac{\delta}{\delta A_\mu} \det_{ren}(1+K_i(A)) = 0$$

where $D_\mu(A)$ is the covariant derivative.

3) By generalizing these determinants, coupling external sources to axial vector currents, it is also possible to derive the usual Adler anomalies [53,70,71].

4) A related fact is the equivalence of the parameter θ introduced in the fermion action in Section 1 with the θ of the θ-states defined by inserting $\exp \frac{i\theta}{2\pi} \int F$ for QED_2 or $\exp \frac{i}{8\pi^2} \int \mathrm{Tr}\, F \wedge F$ for QCD_4 into the measure.

This follows from

Theorem 6.13' : Let $A^\varepsilon \to A$ in a topology sufficiently strong for $\det_{ren}$ to converge; assume furthermore that A is a continuum gauge field with topological charge $n \in \mathbb{Z}$ (i.e. $\int F = 2\pi n$ for QED_2 or $\mathrm{Tr} \int F \wedge F = 8\pi^2 n$ for QCD_4) . Then

$$e^{i\theta n}\det_{ren}(1+K_F(A)) = \lim_{\varepsilon\to 0} \det_{ren}(1+K^{\varepsilon}_{\theta,F}(A^{\varepsilon}))$$

where $K^{\varepsilon}_{\theta,F}(A^{\varepsilon})$ is defined like K^{ε}_F but using an arbitrary angle θ in the lattice fermion action (cf. Section 1) instead of $\theta = 0$.

For a proof see [71].

We promised in the previous subsection that determinants would help to prove convergence of fermion Green's functions. This is easy if we use a little lemma about Lipschitz continuity of Green's functions times determinants :

Lemma 6.13 [34] : $(1+K)^{-1}\det_p(1+K)$ and more generally

$$\Lambda^k((1+K)^{-1})\det_p(1+K)$$

are Lipschitz continuous on I_p where on the image space we use norm topology.

Remark : $\Lambda^k((1+K)^{-1})$ is the operator induced by $(1+K)^{-1}$ on the k-fold antisymmetric tensor product of the underlying Hilbert space H_F .

Matthews and Salam [14] gave already the (obvious) formulae for Fermion Green's functions :

$$G'_F(x,y;A) \equiv ((p^2+M^2)^{-1/4}(1+K_F(A))^{-1} \frac{-i\not{p}+M}{(p^2+M^2)^{3/4}})(x,y) \tag{6.28}$$

$$\begin{aligned} &G'_F(x_1,\dots,x_k;y_1,\dots,y_k;A) = \\ &= \Lambda^k[(p^2+M^2)^{-1/4}(1+K_F(A))^{-1} \frac{-i\not{p}+M}{(p^2+M^2)^{3/4}}](x_1,\dots,x_k;y_1,\dots,y_k) \end{aligned} \tag{6.29}$$

where $G_F = (i\not{p}+M)^{-1}$.

So from Lemma 6.13 and the construction of $\det_{ren}$ we get :

Theorem 6.14 : Let $A^{\varepsilon} \to A$ in such a way that the limit of the lattice determinants exists and is nonzero. Then the lattice fermion Green's functions

$$G^{\varepsilon\,\prime}_F(x_1,\dots,x_k;y_1,\dots,y_k;A^{\varepsilon})$$

also converge in the sense of the norm of operators on $\Lambda^k(H_F)$ $(p > d)$.

Proof : Obvious by the remarks made above. □

c) <u>Convergence of States (Expectation Values) in External Gauge Fields.</u>

For fermions this was just discussed at the end of the last subsection; nothing more has to be done about fermions at this level.

For Bose (Higgs) fields life is much harder because of the self-interaction we have to allow for them.

From now on we will consider the (lattice or continuum) Higgs fields as Gaussian random fields as follows : For each $f \in L^2(\mathbb{R}^2, V_H)$ we have a Gaussian complex random variable $\phi(f)$, depending linearly on f and having mean zero and covariances

$$\begin{aligned}
<\mathrm{Re}\ \phi^{\varepsilon}(f)\mathrm{Re}\ \phi^{\varepsilon}(g)> &= \frac{1}{2}\,\mathrm{Re}(f, C^{\varepsilon}_{g^{\varepsilon}} g) \\
<\mathrm{Im}\ \phi^{\varepsilon}(f)\mathrm{Im}\ \phi^{\varepsilon}(g)> &= \frac{1}{2}\,\mathrm{Re}(f, C^{\varepsilon}_{g^{\varepsilon}} g) \qquad (6.30) \\
<\mathrm{Im}\ \phi^{\varepsilon}(f)\mathrm{Re}\ \phi^{\varepsilon}(g)> &= \frac{1}{2}\,\mathrm{Im}(f, C^{\varepsilon}_{g^{\varepsilon}} g)
\end{aligned}$$

which amounts to

$$\begin{aligned}
<\overline{\phi^{\varepsilon}(f)}\phi^{\varepsilon}(g)> &= (f, C^{\varepsilon}_{g^{\varepsilon}} g) \\
& \qquad\qquad (6.31) \\
<\phi^{\varepsilon}(f)\phi^{\varepsilon}(g)> &= 0
\end{aligned}$$

(remember the identification of $\ell^2(\varepsilon\,\mathbb{Z}^d)$ with a subspace of $L^2(\mathbb{R}^d)$) .

Furthermore we write, after picking an orthonormal basis $\{e_a\}$ in V_H

$$\phi^{\varepsilon}_a(x) \equiv \phi^{\varepsilon}(\delta^{\varepsilon}_{x,a}) \qquad (6.32)$$

where $\delta^{\varepsilon}_{x,a}$ takes the value e_a at x and zero at all other points of $\varepsilon\,\mathbb{Z}^d$ (here we have to assume $\varepsilon > 0$).

Next we define Wick ordered (with respect to the free measure) monomials in ϕ^{ε} by

$$\begin{aligned}
:e^{\lambda \mathrm{Re}\ \phi^{\varepsilon}(f) + \mu\ \mathrm{Im}\ \phi^{\varepsilon}(f)}:_{\varepsilon} &\equiv e^{\lambda\ \mathrm{Re}\ \phi^{\varepsilon}(f) + \mu\ \mathrm{Im}\ \phi^{\varepsilon}(f)} \times \\
& \qquad\qquad (6.33) \\
&\quad \times e^{-1/4(\lambda^2+\mu^2)\mathrm{Re}(f, C^{\varepsilon}_{\mathbb{1}} f)}
\end{aligned}$$

and Taylor expansion. In particular we are interested in

$$:|\phi^{\varepsilon}(f)|^{2n}:_{\varepsilon} \equiv :(\mathrm{Re}\ \phi(f)^2+\mathrm{Im}\ \phi(f)^2)^n:_{\varepsilon} \tag{6.34}$$

and

$$:|\phi^{\varepsilon}(x)|^{2n}:_{\varepsilon} \equiv :(\sum_a (\mathrm{Re}\ \phi_a^{\varepsilon}(x)^2+\mathrm{Im}\ \phi_a^{\varepsilon}(x)^2))^n:_{\varepsilon} \tag{6.35}$$

Notice the free covariance occurring in the definition (6.33).

We denote the Gaussian probability measure corresponding to the random fields ϕ^{ε} by $d\nu^{\varepsilon}_{g^{\varepsilon}}(\phi)$, i.e. $d\nu^{\varepsilon}_{g^{\varepsilon}}$ is specified by

$$\int \phi^{\varepsilon}(f)d\nu^{\varepsilon}_{g^{\varepsilon}} = \int \overline{\phi^{\varepsilon}(f)d\nu^{\varepsilon}_{g^{\varepsilon}}} = 0 \tag{6.36}$$

and

$$\int \overline{\phi^{\varepsilon}(f)}\phi^{\varepsilon}(g)d\nu^{\varepsilon}_{g^{\varepsilon}} = (f,C^{\varepsilon}_{g^{\varepsilon}}g) \quad . \tag{6.37}$$

From Subsection a) we see immediately that the moments and characteristic functions of $d\nu^{\varepsilon}_{g^{\varepsilon}}$ converge as $\varepsilon \to 0$; this is the analogue of the result for fermions mentioned above.

Our selfinteractions are given in terms of a "potential" V which is an even polynomial of degree at least 4 and positive leading coefficient :

$$V(r) \equiv \sum_{k=o}^{p} a_{2k}\ r^{2k} \tag{6.38}$$

Let χ_{Λ} be the characteristic function of a bounded open domain $\Lambda \subset \mathbb{R}^d$. Then we define the "interactions in the volume Λ" by

$$V^{\varepsilon}_{\Lambda}(\phi) \equiv \sum_x \varepsilon^d \sum_{k=o}^{p} a_{2k}\ :|\phi^{\varepsilon}(x)|^{2k}:_{\varepsilon}\ \chi_{\Lambda}(x) \quad . \tag{6.39}$$

After all these definitions it is time for a theorem :

Theorem 6.15 [28] : Let $d = 2$; assume that the gauge fields g^{ε} converge to a continuum gauge field in the sense of Definition 6.4. Then the probability measures

$$\frac{1}{Z^{\varepsilon}_{\Lambda}(g^{\varepsilon})}\, e^{-V^{\varepsilon}_{\Lambda}(\phi)}\, d\nu_{g^{\varepsilon}}(\phi) \tag{6.40}$$

where

$$Z^{\varepsilon}_{\Lambda}(g^{\varepsilon}) \equiv \int e^{-V^{\varepsilon}_{\Lambda}(\phi)}\, d\nu^{\varepsilon}_{g^{\varepsilon}}(\phi) \tag{6.41}$$

converge as $\varepsilon \to 0$ in the sense of characteristic functions; all moments converge. The limit is independent of the orientation of the lattices.

Remarks :

1) Here is is convenient to "index" the random Higgs fields by functions in $S(\mathbb{R}^2,V_H)$ (Schwartz space) instead of L^2 .

2) Using the technology developed for proving stability of ϕ_3^4 by Glimm and Jaffe [35] and others (see for instance [10]) and the ingredients assembled here, it should be possible to prove this theorem also for the three dimensions. Part of this task has been accomplished by Potthoff [31]; the additional problems due to the different Wick ordering should be manageable according to the discussions in Subsection a).

Proof of Theorem 6.15 : We break up the proof into several steps. The first and biggest one is

Lemma 6.16 : Under the assumptions of Theorem 6.15

$$\lim_{\varepsilon \to 0} \int d\nu^{\varepsilon}_{g^{\varepsilon}} e^{-\lambda V^{\varepsilon}_{\Lambda}}$$

exists for all $\lambda \geq 0$.

Proof : We may assume $\lambda = 1$. A helpful trick is to consider all the fields ϕ^{ε} for different ε as being defined on the same underlying measure space; in fact we will consider them as functions of a fixed white noise field as follows :

Let ψ be a V_H-valued white noise field on $\mathbb{R}^2$ and dw the corresponding probability measure, i.e. for $f,g \in S(\mathbb{R}^2,V_H)$

$$\int dw\, \overline{\psi}(f)\psi(g) = (f,g)_{L^2}$$
$$\int dw\, \psi(f)\psi(g) = 0 \quad , \text{etc.} \tag{6.42}$$

Let Q^{ε} be the averaging operator mapping L^2 onto ℓ^2 as discussed earlier, and

$$E^{\varepsilon} \equiv (Q^{\varepsilon *} \chi_{\Lambda} C^{\varepsilon}_{g^{\varepsilon}} \chi_{\Lambda} Q^{\varepsilon})^{1/2} \quad . \tag{6.43}$$

Then we recognize

$$\phi^{\varepsilon} = E^{\varepsilon}\psi \tag{6.44}$$

which is symbolic notation for

$$\phi^{\varepsilon}(f) = \psi(E^{\varepsilon}f) \quad . \tag{6.45}$$

We can thus consider $V_{\Lambda}^{\varepsilon}(\phi^{\varepsilon})$ as a function $\tilde{V}_{\Lambda}^{\varepsilon}(\psi)$ of ψ and $\int d\nu_{g^{\varepsilon}}^{\varepsilon}\exp(-\lambda V_{\Lambda}^{\varepsilon}) = \int dw\, \exp(-\lambda\tilde{V}_{\Lambda}^{\varepsilon})$.

So to show convergence we estimate

$$\begin{aligned} \left| \int dw(e^{-\tilde{V}_{\Lambda}^{\varepsilon}}-e^{-\tilde{V}_{\Lambda}^{\varepsilon'}})\right| &\leq \int dw|\tilde{V}_{\Lambda}^{\varepsilon}-\tilde{V}_{\Lambda}^{\varepsilon'}| \quad (e^{-\tilde{V}_{\Lambda}^{\varepsilon}}+e^{-\tilde{V}_{\Lambda}^{\varepsilon'}}) \leq \\ &\leq (\int dw|\tilde{V}_{\Lambda}^{\varepsilon}-\tilde{V}_{\Lambda}^{\varepsilon'}|^2)^{1/2}\{(\int dw\, e^{-2\tilde{V}_{\Lambda}^{\varepsilon}})^{1/2} + (\int dw\, e^{-2\tilde{V}_{\Lambda}^{\varepsilon'}})^{1/2}\} . \end{aligned} \tag{6.46}$$

First we use diamagnetism to bound the terms in curly brackets : By Theorem 2.6

$$\begin{aligned} \int dw\, e^{-2\tilde{V}_{\Lambda}^{\varepsilon}}\det(1+K_H^{\varepsilon}(A^{\varepsilon}))^{-1} &= \int d\nu_{g^{\varepsilon}}^{\varepsilon} e^{-2V_{\Lambda}^{\varepsilon}}\det(1+K_H^{\varepsilon}(A^{\varepsilon}))^{-1} \leq \\ &\leq \int d\nu_{\mathbb{1}}^{\varepsilon}\, e^{-2V_{\Lambda}^{\varepsilon}} \end{aligned} \tag{6.47}$$

(the left hand side is just $Z(\{g_{xy}\}))$. A uniform (in ε) bound on the right hand side of (6.47) has first been given by Nelson [36] and has long since become for constructive field theorists as basic as Schwarz's inequality (see also [11]). The determinant we had to throw into (6.47) has been controlled in the previous subsection, so we are done with the curly brackets.

To see convergence to zero of the first factor in (6.46), it is useful to rewrite $V_{\Lambda}^{\varepsilon}$ as a sum of monomials Wick ordered with respect to $C_{g^{\varepsilon}}^{\varepsilon}$ (these are defined by writing $C_{g^{\varepsilon}}^{\varepsilon}$ in (6.33) in the place of $C_{\mathbb{1}}^{\varepsilon}$) . This has been worked out explicitly in [37], for instance, but we only need to know that this produces coefficients that are polynomials in $\delta C_{g^{\varepsilon}}^{\varepsilon}$ which was discussed in 6.a). So it suffices to show

$$\lim_{\varepsilon,\varepsilon'\to 0} \int dw|P^{\varepsilon}(\phi^{\varepsilon})-P^{\varepsilon'}(\phi^{\varepsilon'})|^2 = 0 \tag{6.48}$$

where

$$P^{\varepsilon} \equiv \int_{\Lambda} |\delta C_{g^{\varepsilon}}^{\varepsilon}(x,x)|^j :|\phi^{\varepsilon}|^{2N}:_{C_{g^{\varepsilon}}^{\varepsilon}} dx \tag{6.49}$$

(6.47) may be further reduced to proving

$$\int dw\, P^{\varepsilon}(P^{\varepsilon}-P^{\varepsilon'}) \to 0 \tag{6.50}$$

which in turn will follow from

$$\begin{aligned}\int_{\Lambda\times\Lambda} dx\, dy |\delta C^{\varepsilon}_{g^{\varepsilon}}(x,x)|^{j}[(E^{\varepsilon})^{2N}(x,y)|\delta C^{\varepsilon}_{g^{\varepsilon}}(y,y)|^{j} - \\ - (E^{\varepsilon}E^{\varepsilon'})^{N}(x,y)|\delta C^{\varepsilon'}_{g^{\varepsilon'}}(y,y)|^{j}]\end{aligned} \tag{6.51}$$

by the rules of Gaussian integration.

After telescoping and using a few Hölder inequalities this reduces to showing $L^{p}(\Lambda \times \Lambda)$ convergence of $(E^{\varepsilon}E^{\varepsilon'})(x,y)$ (remember Theorem 6.6 showing L^{p} convergence of $\delta C^{\varepsilon}_{g^{\varepsilon}}$) . Now

$$\begin{aligned}&\int |(E^{\varepsilon}E^{\varepsilon'})(x,y)|^{2p} dx\, dy \leq \\ &\leq (\int |(E^{\varepsilon})^{2}(x,y)|^{2p} dx\, dy)^{1/2} (\int |(E^{\varepsilon'})^{2}(x,y)|^{2p} dx\, dy)^{1/2}\end{aligned}$$

which is uniformly bounded by diamagnetism. So by Hölder's inequality it suffices to show $L^{2}(\Lambda \times \Lambda)$ convergence of the kernel of $E^{\varepsilon}E^{\varepsilon'}$, which means I_2 convergence of the operators $E^{\varepsilon}E^{\varepsilon'}$; this in turn follows from I_4 convergence of E^{ε} .

Most of Subsection 6.a) was spent proving things like I_2 convergence of $(E^{\varepsilon})^{2}$. Now it is a fact (proven in [28]) that this implies the desired I_4 convergence of E^{ε} . So Lemma 6.16 is true. □

Next we have to show

<u>Lemma 6.17</u> :

$$\lim_{\varepsilon\to 0} \int d\nu^{\varepsilon}_{g^{\varepsilon}}\, e^{-V^{\varepsilon}_{\Lambda}} > 0 \quad .$$

<u>Proof</u> : By the convexity of the exponential (Jensen's inequality) it is enough to show

$$\int d\nu^{\varepsilon}_{g^{\varepsilon}} V^{\varepsilon}_{\Lambda} < C < \infty$$

where C is independent of ε . This is easy by explicit Gaussian integration. □

To complete the proof of Theorem 6.15 consider

$$\int F(\phi^{\varepsilon})e^{-V_{\Lambda}^{\varepsilon}}d\nu_{g^{\varepsilon}}^{\varepsilon} \tag{6.52}$$

where F is a polynomial or an exponential. Convergence of (6.52) follows from Lemma 6.16 and convergence of

$$\int |F(\phi^{\varepsilon})-F(\phi^{\varepsilon'})|^2 d\nu_{g^{\varepsilon}}^{\varepsilon} \tag{6.53}$$

which is true because Gaussian integration reduces it to convergence of Green's functions. So we are done with Theorem 6.15.

□

We have obtained, after some suffering, the continuum state (i.e. expectation value) for two-dimensional fermion or Higgs fields coupled to an external Yang-Mills field with some regularity (i.e. Hölder continuity which is essentially equivalent to finiteness of $\|A\|_{\infty,\alpha}$).

d) Convergence of Expectations in Fully Quantized Theories with Cutoff on the Gauge Field.

The results of the previous subsection easily extend to convergence in fully quantized theories provided the gauge field measure is such that the gauge fields are Hölder continuous with probability 1. The trouble is :

1) The correct Yang-Mills measure (if we can lay our hands on it) certainly will not have this property, so we will have to cut it off in some way.

2) It is hard to find cutoffs that do not destroy most of the structure of a Yang-Mills theory - this was the motivation in the beginning for working with a lattice as a cutoff. But for abelian fields there are reasonable continuum cutoffs with the right properties and we will use these.

Theorem 6.18 : Let $d = 2$ and dm be a probability measure defining a random Yang-Mills field A_μ that is essentially uniformly Hölder continuous with some index $\alpha > 0$ with probability one. Let $\{g_{xy}\}$ be the lattice gauge field induced by A_μ. Then the probability measures

$$\frac{1}{Z_{\varepsilon,\Lambda}} dm \times e^{-V_{\Lambda}^{\varepsilon}} \det_{ren}(1+K_H^{\varepsilon}(A^{\varepsilon}))^{-1} d\nu_{g^{\varepsilon}}^{\varepsilon}(\phi)$$

converge as $\varepsilon \to 0$ in the sense of characteristic functions and moments where $Z_{\varepsilon,\Lambda}$ is the appropriate normalization.

Remarks :

1) Essentially uniformly Hölder continuous means : For m-almost all A_μ there is a constant c_A such that

$$|A_\mu(x)-A_\mu(y)| \le c_A |x-y|^\alpha , \quad \mu = 0,1$$

for (Lebesgue-) almost all x,y .

2) Assuming the generalization of Theorem 3.15 to three dimensions this theorem also would generalize to d = 3 .

There is a fermionic analogue :

Theorem 6.19 : Let dm be as in the previous theorem and d = 2 or 3 . Then the probability measure

$$\frac{1}{Z_{\varepsilon,\Lambda}} \det_{ren}(1+K_F^\varepsilon(A^\varepsilon))dm$$

as well as the measures

$$\frac{1}{Z_{\varepsilon,\Lambda}} \Lambda^k(G_F^\varepsilon(1+K_F^\varepsilon(A^\varepsilon))^{-1})\det_{ren}(1+K_F^\varepsilon(A^\varepsilon))$$

(that take values in $L(\Lambda^k(H_F))$) converge as $\varepsilon \to 0$ in the sense of characteristic functions and moments.

Proof of both theorems : This is just Theorem 6.15 (and its trivial fermionic analogue) together with the dominated convergence theorem. The uniform upper bound comes from "diamagnetism" which throws out the coupling between matter and gauge fields.

We should stress once more that the two theorems sound better than they are since for nonabelian theories we do not have the required measures. But for abelian theories in d = 2,3 such measures are found easily, due to the following theorem due to A.M. Garsia [79] :

Theorem 6.20 : Let $\phi(x)$ be a Gaussian random field on a bounded region Λ . A sufficient condition for its sample paths to be almost surely essentially uniformly Hölder continuous with index α is that the function

$$p(u) \equiv \sup_{|x-y| \le |u|} \langle(\phi(x)-\phi(y))^2\rangle^{1/2}$$

be Hölder continuous with index $\beta > \alpha$ at $u = 0$.

We refer to [79] for a proof. The use of this lies in the following

Corollary 6.21 [28] : Let d be arbitrary, and let A_ε be the d-component Gaussian random field with covariance

$$<A_\mu(x)A_\nu(y)> \equiv D_{\mu\nu}(x,y)$$
$$\sum_\mu \int dk\ \hat{D}_{\mu\mu}(k)|k|^{2\beta} < \infty \tag{6.54}$$

Then with probability one A_μ is essentially uniformly Hölder continuous of any index $\alpha < \beta$.

Proof : A simple exercise in Fourier transformation.

□

Example :

$$\hat{D}_{\mu\nu}(k) = (\delta_{\mu\nu} - \frac{k_\mu k_\nu}{k^2+\lambda^2}) \frac{1}{k^2+\lambda^2} e^{-t\vec{k}^2} \quad , \quad t > 0 \tag{6.55}$$

To check (6.54) for this covariance for any $\beta < \frac{1}{2}$ is trivial. (6.55) is the covariance of a "photon field" with ultraviolet cutoff t only on the spatial momenta, infrared cutoff λ and some gauge fixing close to the Landau gauge. The main advantage of the peculiar momentum cutoff is that it preserves Osterwalder-Schrader positivity :

Theorem 6.22 [28] : Let an expectation on a combined gauge field - matter field system be given by either

$$\lim_{\varepsilon\to o} \frac{1}{Z_{\varepsilon,\Lambda}} e^{-V^\varepsilon_\Lambda} \det_{ren}(1+K^\varepsilon_H(A^\varepsilon))^{-1}\, d\nu^\varepsilon_{g^\varepsilon}(\phi)$$

for a two-dimensional abelian Higgs model or by the measures given in Theorem 6.19 for two or three-dimensional QED . Then the expectation values are Osterwalder-Schrader positive provided Λ is symmetric about $t = 0$.

Proof : This follows essentially from O.S. positivity on the lattice by taking limits. To be on safe ground one has to introduce for a moment lattice cutoffs ε' for the gauge field and ε for the matter fields and one sends first ε' and then ε to zero. See [28].

□

This concludes our study of "nice" gauge fields; what remains is to remove the ultraviolet cutoff on the gauge field and eventually the volume cutoff and to check the axioms of Osterwalder and Schrader.

7. REMOVAL OF ALL CUTOFFS; VERIFICATION OF AXIOMS IN TWO DIMENSIONS

The ultraviolet cutoff on the gauge field has been removed in [29] for 2-dimensional Higgs models and in [26,27] for QED_2 ; for the latter case I am not aware of any detailed published study of the infinite volume limit and verification of the axioms which have been carried out in [29] for Higgs matter. There is, however, no obstacle in principle even in the fermionic case - actually it should be easier - and we will mostly discuss both situations in parallel. For the case of QED_2 I will give an outline of a construction of the thermodynamic limit by means of a cluster expansion whereas for $Higgs_2$ a simple argument based on correlation inequalities can be used.

a) The Stability Expansion

To send the t-cutoff inherent in the gauge field measure (cf. 6.55)) to zero it is convenient to use a so-called stability expansion (actually in QED_2 this can be replaced by a simpler argument [27] based on [15]). We discuss the Higgs model and comment along the way on the appropriate modifications (mostly simplifications) for QED_2 . The idea is simply to choose a suitable sequence of cutoffs $t_0, t_1, t_2, \ldots \to 0$, telescope (modified) partition functions

$$Z(t_N) = Z(t_o) + \sum_{k=1}^{N} (Z(t_k) - Z(t_{k-1})) \tag{7.1}$$

and estimate the differences by convergent Feynman graphs. Of course this will only work if the appropriate counterterms are inserted.

To be more specific, let us fix some notation. We denote by

$$dm_t(A)$$

the Gaussian measure on two-component fields A_μ with mean zero and covariance $D^t_{\mu\nu}$ given by

$$\hat{D}^t_{\mu\nu} \equiv (\delta_{\mu\nu} - \frac{k_\mu k_\nu}{k^2+\lambda^2})(k^2+\lambda^2)^{-1}\, e^{-tk_1^2} \qquad (\mu,\nu = 0,1) \tag{7.2}$$

(cf. (6.55)). We also need Gaussian measures with covariances

$$D_{\mu\nu}^{t_k} - D_{\mu\nu}^{t_{k-1}} \tag{7.3}$$

assuming $\infty = t_0, t_1, t_2, \ldots$ is a monotonically decreasing sequence. We call the measures corresponding to (7.3) $dm^{(k)}(A)$, $k = 1,2,\ldots$.

We can identify $dm_{t_N}(A)$ with the product measure $\prod_{k=1}^{N} dm^{(k)}(A)$ and $dm(A)$ with dm_{t_∞} .

We will also use some interpolating fields to estimate the differences in (7.1), namely if $A_\mu^{(k)}$ is the random field corresponding to the measure $dm^{(k)}(A)$, we define

$$A_{\mu,s(\ell)} \equiv \sum_{i=1}^{\ell} \sqrt{s_1 \ldots s_i}\, A_\mu^{(i)} \tag{7.4}$$

$$(s_i \in [0,1] \quad ; \quad i = 1,2,3,\ldots) \quad .$$

This strange looking interpolation is chosen to make the interpolating covariances simple.

There will be a vacuum energy counterterm

$$E_\Lambda^{(t)} \equiv e^2 \int_{\Lambda\times\Lambda} dx\, dy \int A_\mu(x) A_\nu(y) \times \Pi_{\mu\nu}(x-y)\, dm_t(A) \tag{7.5}$$

where $\Pi_{\mu\nu}$ is the second order vacuum polarization. Graphically

$\Pi_{\mu\nu}$

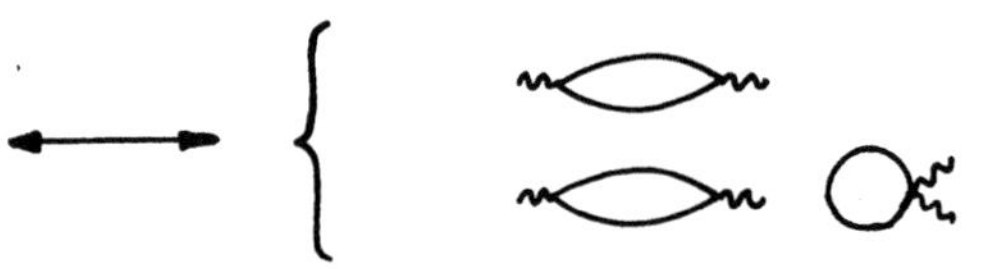

in QED_2

in $Higgs_2$

and

E_Λ^t

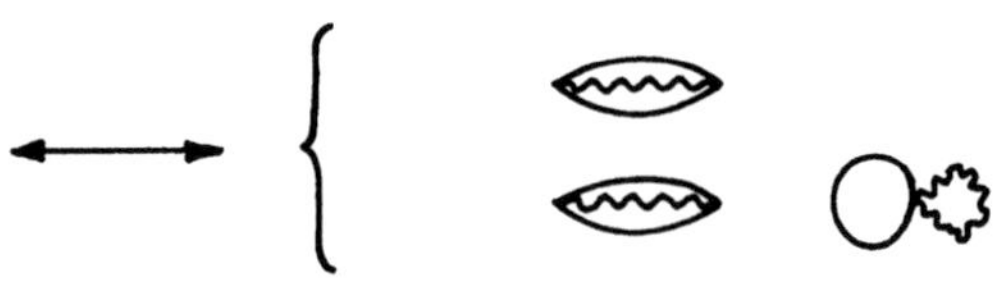

in QED_2

in $Higgs_2$

Only in $Higgs_2$ there will be a mass counterterm for the matter field which may be chosen for our gauge fixed covariance to be

$$\delta m_t^2 = \sum_\mu e^2 \int A_\mu(0)^2 \, dm_t(A) = e^2 \sum_\mu D_{\mu\mu}^t(0) \quad .$$

Graphically

$\delta m_t^2 =$  in Higgs_2 .

Interpolated counterterms $E_{\Lambda,s(\ell)}$, $\delta m^2_{s(\ell)}$ are defined accordingly.

Finally we denote by $d\nu_A(\phi)$ the continuum Gaussian probability measure with mean zero and covariance $C_A(x,y) \equiv (-\Delta_A + m^2)^{-1}(x,y)$ and define

$$z(A) \equiv \begin{cases} \det_{\text{ren}}(1+K_H(A))^{-1} & \text{in Higgs}_2 \\ \\ \det_{\text{ren}}(1+K_F(A)) & \text{in QED}_2 \end{cases}$$

The full unnormalized cutoff measure for Higgs_2 is

$$Z_{\Lambda,t} d\mu_{\Lambda,t}(\phi,A) \equiv dm_t(A) z(A) e^{-V_\Lambda(\phi) + \frac{1}{2}\delta m_t^2 \int_\Lambda :|\phi|^2: dx + E^t_\Lambda} d\nu_A(\phi) \qquad (7.6)$$

where $V_\Lambda(\phi) = \lim_{\varepsilon \to o} V^\varepsilon_\Lambda(\phi)$ (see the previous subsection). For QED_2 one has instead to consider the "$L(H^k(\Lambda_F))$ valued measures"

$$dm_t(A) z(A) \Lambda^k (G_F(1+K_F(A))^{-1}) \quad . \qquad (7.7)$$

We denote by $Z_{\Lambda,t}$ ("partition function") the integrals of (7.6) and (7.7), respectively, and by $Z_{P,\Lambda,t}$ modified partition functions (unnormalized expectations).

If P is a polynomial in Higgs and gauge fields

$$Z_{P,\Lambda,t} \equiv \int P \, d\mu_{\Lambda,t} \quad . \qquad (7.8)$$

For QED_2 it is convenient to think of an underlying Grassmann "measure" generating the Matthews-Salam formulas. P is then a polynomial in gauge and fermion fields :

$$P = \sum_k \int B_k(x_1,\ldots,x_k;y_1,\ldots,y_k;A) \prod_{i=1}^k (\bar\psi(x_i)\psi(y_i))$$

and

$$Z_{P,\Lambda,t} \equiv \int dm_t(A) z(A) \times \sum_k \operatorname{Tr}[B_k \Lambda^k (G_F(1+K_F(A))^{-1})] \quad . \qquad (7.9)$$

Next we use the interpolating fields, the fundamental theorem of calculus and integration by parts to obtain an expression for the difference $Z_{P,\Lambda,t_k} - Z_{P,\Lambda,t_{k-1}}$.

Lemma 7.1 : For Higgs_2

$$Z_{P,\Lambda,t_k} - Z_{P,\Lambda,t_{k-1}} = \int_0^1 ds_1 \dots \int_0^1 ds_k \int d\mu_{s(k)}(A,\phi) K_k \dots K_1 P$$

where $K_1,\dots,K_k$ are functional differential operators acting on P . They can be represented graphically as

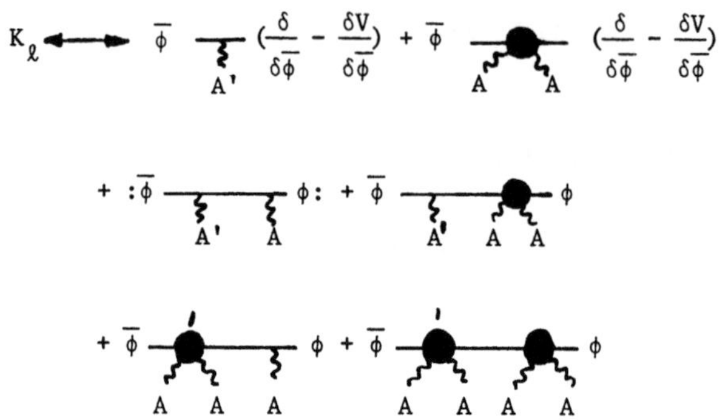

+ $:A_\mu \Pi_{\mu\nu} A_\nu:$

(7.10)

Remark : We hope that the graphical notation is to a large extent self-explanatory to anyone who is familiar with Feynman graphs. The action of functional derivatives can be defined purely algebraically since they only act on polynomials (or possibly exponentials of polynomials). No sophisticated functional analysis is involved. A detailed explanation of the notation is given in [28].

A' always stands for $\frac{\delta}{\delta s_\ell} A_{s(\ell)}$, the black box  stands for 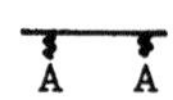 - $\delta m^2_{s(\ell)}$ and  for its derivative with respect to s_ℓ . —— stands of course for the free two point function of the Higgs field.

Sketch of proof : (see [28] for details) : We claim that for $\ell < k$

$$Z_{P,\Lambda,t_k} - Z_{P,\Lambda,t_{k-1}} = \int_0^1 ds_1 \dots \int_0^1 ds_\ell [\int d\mu_{(s_1,\dots,s_\ell,1,\dots,1)}(A) - $$

$$- \int d\mu_{(s_1,\dots,s_\ell,1,\dots,1,0)}] K_\ell \dots K_1 P$$

This is proven by induction. It is trivial for $\ell = 0$; to go from ℓ to $\ell+1 < k$ we write the difference of the measures appearing in (7.11) as

$$\int_0^1 ds_{\ell+1} \frac{\partial}{\partial s_{\ell+1}} (d\mu_{(s_1,\dots,s_{\ell+1},1,\dots,1)} - d\mu_{(s_1,\dots,s_{\ell+1},1,\dots,1,0)}) \tag{7.12}$$

which is true because the difference of measures in (7.12) vanishes at $s_{\ell+1} = 0$ due to the definition (7.4) of interpolating fields.

The derivative in (7.12) can be worked out :

$$\frac{\partial}{\partial s_{\ell+1}} d\mu(s_1,\ldots,s_{\ell+1},\ldots) = (E'_\Lambda + \bar{\phi} \underset{A}{\text{——}} \phi + \bar{\phi} \underset{A \quad A}{\text{—●—}} \phi) d\mu(s_1,\ldots,s_{\ell+1},\ldots) \tag{7.13}$$

by differentiating the action. The rest is an exercise in integration by parts in Gaussian integrals; this contracts the fields occuring in (7.13) with fields in the action. This process is stopped when all the expressions occurring downstairs give only rise to convergent Feynman integrals when contracted with themselves using the free Gaussian measure. This is how $K_{\ell+1}$ is produced and (7.11) for $\ell+1$ is proven.

Finally in the last step $\ell = k$ the expression quoted in the lemma appears.

For a careful justification of the integration by parts see [28] where this is done with the help of the lattice approximation.

There is a fermionic analogue to Lemma 7.1 :

Lemma 7.2 : For QED_2

$$Z_{P,\Lambda,t_k} - Z_{P,\Lambda,t_{k-1}} = \int_0^1 ds_1 \ldots \int_0^1 ds_k \int d\mu_{s(k)}(A) K_k \ldots K_1 P$$

where $K_k,\ldots,K_1$ are some (non-linear) operators acting on P . Again they are best represented graphically and it makes them more transparent if they are expressed with respect to the underlying fermion expectation :

$$K_\ell \longleftrightarrow : \bar{\psi} \underset{A' \quad A}{\text{————}} \psi : + \bar{\psi} \underset{A'}{\text{——}} \frac{\delta}{\delta \bar{\psi}} + :A_\mu \Pi_{\mu\nu} A_\nu: \tag{7.14}$$

Proof : Completely analogous to the proof of Lemma 7.1, only simpler. Integration by parts with respect to the Berezin "integration" is discussed in [74].

□

From the two lemmas we get

Corollary 7.3 : For $Higgs_2$ or QED_2

$$Z_{P,\Lambda,t_N} = \sum_{\ell=1}^{N} \int_0^1 ds_1 \ldots \int_0^1 ds \, d\mu_{s(\ell)} K_\ell \ldots K_1 P \qquad (7.15$$

with $K_r (r \in \mathbb{Z}_+)$ given by (7.10) for Higgs_2 and by (7.14) for QED_2 .

Remark : We call (7.15) the stability expansion.

The point of (7.15) is of course that one can show uniform convergence and thereby take the limit $N \to \infty$, provided the cutoff sequence is chosen appropriately.

Let us first talk about the Higgs case. The necessary estimates are highly plausible (even though the detailed proof takes up 20 pages in the preprint of [29]).

Lemma 7.4 : In Higgs_2

$$|Z_{P,\Lambda,t_k} - Z_{P,\Lambda,t_{k-1}}| \leq C_1 |\log t_k|^{kr} \prod_{j=1}^{k} t_j^{\delta} (k!)^p \, e^{C_2 (\log t_k)^2} \tag{7.16}$$

for some constants $C_1, C_2, \delta, r, p > 0$.

Corollary 7.5 : Let $t_j = \text{const}\, e^{-j^{\gamma}}$ $(0 < \gamma < 1)$ for $j = 1,2,3,\ldots$. Then (7.16) implies absolute and uniform convergence of the stability expansion (7.15).

Proof of Corollary 7.5 : If $t_j = \exp(-j^{\gamma})$

$$|Z_{P,\Lambda,t_k} - Z_{P,\Lambda,t_{k-1}}| \leq$$

$$\leq C_1 k^{\gamma k r} \exp(-\varepsilon \sum_{j=1}^{k} j^{\gamma}) k^{pk} \exp(C_2 k^{2\gamma})$$

$$\leq C_1 \exp\{(p+r\gamma)k \log k + C_2 k^{2\gamma} - \frac{\varepsilon}{\gamma+1} k^{\gamma+1}\}$$

$$\leq C_1 \exp(-C_3 k^{\gamma+1})$$

for some $C_3 > 0$. This is clearly summable over k .

□

To proceed with the proof of Lemma 7.4 one first gets rid of the remaining exponential by Schwarz's inequality and diamagnetism :

$$| \int d\nu_{A_{s(\ell)}} (\phi) z(A)\, e^{-V_\Lambda + \frac{1}{2} \delta m^2_{s(\ell)} \int_\Lambda :|\phi|^2: + E_{s(\ell)}} K_\ell \ldots K_1 P| \leq \tag{7.17}$$

$$\leq (\int d\nu_{A_{s(\ell)}} |K_\ell \ldots K_1 P|^2)^{1/2} (\int d\nu_{A_{s(\ell)}} e^{-2V_\Lambda} z(A) e^{\delta m^2_{s(\ell)} \int_\Lambda :|\phi|^2:})^{1/2} e^{E_{s(\ell)}}$$

Lemma 7.4 will be a consequence of the following three lemmas and (7.17).

Lemma 7.6 : $E(s_1,\ldots,s_\ell,0,\ldots) \leq a_1(\log t_\ell)^2$.

Lemma 7.7 : $\int d\nu_A z(A) e^{-2V_\Lambda+\delta m^2\int_\Lambda :|\phi|^2:} \leq \exp a_2(\delta m^2)^2$ provided V contains a term $\lambda\int:(\bar{\phi}\phi)^2:$, $\lambda > 0$ (which we will assume for simplicity). Furthermore

$$\delta m^2_{s(\ell)} \leq a_3|\log t_\ell| \quad .$$

Lemma 7.8 : $\int dm(A) \int d\nu_A |K_\ell\ldots K_1 P|^2 \leq a_4^\ell(\prod_{j=1}^{\ell} t_j^\delta)(\ell!)^p|\log t_\ell|^{\ell r}$ for some $\delta > 0$, $p > 0$, $r > 0$.

Remark : The assumption in Lemma 7.7 about a positive 4th order term is not really necessary; if the highest order term in V is $\lambda:(\bar{\phi}\phi)^{2N}:$, $\lambda > 0$ we always get a bound of the form $\exp a_2(\delta m^2)^{2N/2N-2}$ which would also give convergence (actually it would make things better).

Proof of Lemma 7.6 : A simple estimate on the Feynman graphs

 □

Proof of Lemma 7.7 : By the diamagnetic bound (Theorem 2.6)

$$\int d\nu_A z(A) e^{-2V_\Lambda+\delta m^2\int_\Lambda :|\phi|^2:} \leq \int d\nu_o \, e^{-2V_\Lambda+\delta m^2\int_\Lambda :|\phi|^2:}$$

So we only have to estimate a Gaussian integral with the free covariance. If it were not for Wick ordering, Lemma 7.7 would follow by taking the maximum of $e^{-2V_\Lambda+\delta m^2\int_\Lambda :|\phi|^2:}$. Because of Wick ordering this is not a bounded function, but the probability of it being large is very small. This is the essence of Nelson's stability proof [36] . More concretely, we write (with $\alpha = \delta m^2$)

$$2V_\Lambda = V_1+V_2 + \alpha\int_\Lambda : |\phi|^2:dx$$

$$V_1 \equiv \lambda(\int_\Lambda :|\phi|^2:dx)^2 - \alpha\int_\Lambda :|\phi|^2:dx$$

$$V_2 \equiv 2\lambda \int_\Lambda :|\phi|^4:dx - \lambda(\int_\Lambda :|\phi|^2:dx)^2 \quad .$$

Now clearly

$$V_1 \geq -\frac{1}{4}\,\frac{\alpha^2}{\lambda}$$

and by Nelson's stability proof [36] for $P(\phi)_2$

$$d\nu_o\, e^{-V_2} < \infty$$

which proves Lemma 7.7.

□

Proof of Lemma 7.8 : This is "just" an estimate of a Gaussian integral that may be expressed as a sum of convergent Feynman diagrams. Unfortunately these Feynman diagrams are arbitrarily large and also the combinatorics is not to be neglected (it is responsible for the factor $(k!)^p$ in (7.16)).

The technique of estimating large Feynman diagrams in terms of a fixed number of "small" ones is quite familiar in constructive field theory. For the case at hand it is carried out in [29] not by a "graphological" but by a functional integral method. The upshot is that

$$\|(\prod_{i=1}^{k} K_i)P\|_2$$

is estimated by $\mathcal{O}((k!)^p)$ terms each of which is a product of $\mathcal{O}(k)$ Feynman diagrams chosen from a certain finite set. Each of these diagrams has at least one differentiated (with respect to a parameter s_i) photon line which gives it a "high momentum", i.e. t values $> t_i$ do not contribute; since we are dealing with power convergent graphs this gives a small factor of the order t_i^δ . Note that the graphs containing the covariant Green's function C_A have to be estimated in terms of those with the free Green's function. This is not entirely trivial because of the derivative coupling, but it is possible (see [29]) and produces the factors $|\log t_k|$.

To extract the power t_i^δ in [29] a general power counting theorem is proven.

Putting everything together gives Lemma 7.8 and so completes the convergence proof of Higgs_2 .

Let us turn for a moment to QED_2 . Here we have

Lemma 7.9 : In QED_2

$$|Z_{P,\Lambda,t_k} - Z_{P,\Lambda,t_{k-1}}| \leq C_1 \prod_{j=1}^{k} t_j^\delta (k!)^p\, e^{C_2|\log t_k|} \tag{7.18}$$

for some constants $C_1, C_2, \delta, p > 0$.

Remark : The factor $(k!)^p$ should not be necessary. The expansion we use is a slight overkill, but since we discussed it for Higgs_2 anyway, we do not mind. In [74] I describe a much simpler stability expansion for QED_2 that requires, however, $e/M \ll 1$.

There is no Schwarz inequality for the fermion "measure" and it will not be needed because there is no self-interaction.

We write $\langle\cdot\rangle_{F,A}$ for the "Gaussian" fermion expectation characterized by

$$\langle\psi(x)\bar{\psi}(y)\rangle_{F,A} = G_F'(x,y;A) \tag{7.19}$$

Then

$$\begin{aligned} &\left|\int dm(A_{s(\ell)})z(A_{s(\ell)})\langle K_\ell\ldots K_1 P\rangle_{F,A}\, e^{E_\Lambda^{s(\ell)}}\right| \\ &\leq \int dm(A_{s(\ell)})\left|\langle K_\ell\ldots K_1 P\rangle_{F,A}\right|^2 e^{E_\Lambda^{s(\ell)}} \end{aligned} \tag{7.20}$$

which is already a sum of Feynman graphs; some of the fermion lines are free (the ones coming from K_i) , others correspond to G_F' . Unfortunately there is no simple way to throw out this nonpolynomial A-dependence other than by using

$$\|(i\partial\!\!\!/+iA\!\!\!/+M)^{-1}\| \leq \frac{1}{M} \tag{7.21}$$

which is possible if the kernel functions $B_k(\underline{x},\underline{y};A)$ appearing in the representation

$$P = \sum_k \int B_k(x_1,\ldots,x_k;y_1,\ldots,y_k;A)\prod_{i=1}^{k}(\bar{\psi}(x_i)\psi(y_i)) \tag{7.22}$$

are sufficiently nice (for instance in $\mathcal{S}$) .

Then one still has to do surgery to cut up the large graphs into small ones; it is possible to use the same bosonic functional methods as for Higgs_2 if one estimates all Feynman graphs involving free fermion lines G_F by graphs where G_F is replaced by the "bosonic" covariance $(-\Delta + m^2)^{-1/2}$. Of course one can also work directly with the graphs. In any case, the bound (7.18) emerges after considerably less work then (7.16).

□

So the stability expansion converges for QED_2 too.

Remarks :

1) In this form the stability expansion certainly does not work in $d = 3$. However

it is plausible that a phase cell expansion (cf. [35], [10]) could be made to work. Magnen and Sénéor [17], skeptical about this possibility, propose instead a cutoff on the matter field that destroys gauge invariance but helps for stability; in the end of course one has to verify that the usual Ward identities are not violated.

2) It is of interest to construct the stability expansion for gauge invariant objects like $P = \overline{\psi}(x)(\exp i \int_x^y A)\psi(y)$. This can be accomodated even though the kernel functions of this P are not in $\mathcal{S}$.

b) Volume Dependent and Volume Independent Bounds.

First we will prove bounds of the form

$$|\log Z_\Lambda| \leq \text{const } |\Lambda| \quad . \tag{7.23}$$

They are essential for controlling the infinite volume limit, whether that is done by a continuum cluster expansion (see [38] and below), or by correlation inequalities [29].

Then by the chessboard bound and closely related techniques that have a long history in constructive field theory (see [40-44], [18]) we will obtain uniform bounds on expectations of exponentials and polynomials in the fields.

Depending on the boundary conditions used, some of the bounds (7.23) become trivial (at least for rectangles Λ).

For instance for free boundary conditions the lower bound contained in (7.23) follows simply from Osterwalder-Schrader positivity : If Λ has sides L,T

$$Z_{L,T} \leq Z_{2L,T}^{1/2} \tag{7.24}$$

(in obvious notation) and therefore

$$\log Z_{L,T}$$

is convex in L and T separately and

$$Z_{L,T} \geq (Z_{\frac{L}{N},\frac{T}{M}})^{NM} \tag{7.25}$$

The upper bound becomes easiest with (anti-) periodic boundary conditions : Formally one has, denoting the (anti-) periodic partition function by $Z_{LT}^{(a)p}$:

$$Z_{LT}^{(a)p} = \mathrm{Tr}\, e^{-TH_L^{(a)p}} \left(\mathrm{Tr}\, e^{-T\overset{o}{H}{}_L^{(a)p}}\right)^{-1} \tag{7.26}$$

where $H_L^{(a)p}$ is the Hamiltonian with (anti-) periodic b.c. and $\overset{o}{H}{}_L^{(a)p}$ the corresponding free Hamiltonian. To prove (7.26) or at least the convexity implied by it one has to go through a lattice approximation; actually the argument is somewhat complicated and requires taking the time continuum limit before the one in space direction. For details see [29]. A different proof of (7.23) for QED_2 has been given by Itô [75].

Note that the left hand side of (7.26) is symmetric in L and T, whereas numerator and denominator on the right hand side are not separately symmetric in L and T. The point of (7.26) if of course that

$$\mathrm{Tr}\, e^{-TH} \leq \left(\mathrm{Tr}\, e^{-\frac{T}{N} H}\right)^N \tag{7.27}$$

and since the denominator in (7.26) can easily be computed explicitly and shown to obey

$$\mathrm{Tr}\, e^{-T\overset{o}{H}{}_L^{(a)p}} \geq e^{-\alpha LT} \tag{7.28}$$

for some $\alpha \in \mathbb{R}$, one can conclude from (7.26) :

$$Z_{LT}^{(a)p} \leq e^{cLT} \tag{7.29}$$

by exploiting the symmetry of $Z_{LT}^{(a)p}$ ("Nelson's symmetry").

To complete the proof of (7.23) one uses the fact that a trace of a positive operator is bigger than any diagonal element to see that the (anti-) periodic partition function essentially dominates all partition functions with other boundary conditions, which can be interpreted as expectation values of e^{-TH} in some (not normalized) states. So

$$Z_{LT}^{X} \leq c_1\, e^{c_2(L+T)} Z_{LT}^{(a)p} \tag{7.30}$$

where X stands for various boundary conditions.

The most useful boundary conditions are :

1) free b.c. (denoted "F") which correspond to turning off the charge of the matter field outside the region in question (this is different from the "free boundary conditions" used in Section 4 for lattice models).

2) Mixed b.c. : 0-Dirichlet b.c. with free Wick ordering for the matter fields, free b.c. for the gauge field (denoted "D_M") . 0-Dirichlet b.c. for fermion fields are a bit tricky (see [39]).

3) Mixed b.c. : Free b.c. for the matter fields, 0-Dirichlet b.c. for the gauge field (denoted "D_G").

To prove (7.30) for these mixed b.c. requires some care because the "states" in which the expectation values of e^{-TH} are taken have infinite norm (like plane waves). See [29] for a detailed treatment of this problem for Higgs_2 . It turns out that to avoid infinities one should require for instance $L,T \geq 1$.

A lower bound for mixed b.c. is obtained in a similar way as for free ones; again one has to be a little careful because of the infinite norm of the "states" appearing there. Formally

$$Z_{LT}^{D_M} = (\eta_L, e^{-TH_L^{D_M}} \eta_L) \times (\overset{o}{\eta}_L, e^{-TH_L^{oD_M}} \eta_L)^{-1} \tag{7.31}$$

where $H_L^{D_M}$, $H_L^{oD_M}$ are the appropriate Hamiltonians and η_L, η_L "infinite norm vectors". Again the denominator is controlled explicitly and one obtains

$$Z_{LT}^{D_M} \geq e^{-cLT} \tag{7.32}$$

We collect all these results in

<u>Theorem 7.10</u> : For $L,T \geq 1$

$$|\log Z_{LT}^X| \leq \text{const } LT \tag{7.33}$$

where in Higgs_2 X can be p , D_M or F and in QED_2 X can be ap , D_M or F .

We will also need upper bounds for modified partition functions. They are easily obtained by the "chessboard bound" (see Theorem 2.2) and related arguments. This works best for unnormalized expectations of exponentials. We define

<u>Definition 7.11</u> : In Higgs_2

$$Z_{g,h;\Lambda}^X \equiv Z_\Lambda^X \langle e^{:|\phi|^2:(g)+F(h)} \rangle_{X,\Lambda}$$

<u>Definition 7.11'</u>: In QED_2

$$Z_{B,h;\Lambda}^X \equiv Z_\Lambda^X \langle e^{(\overline{\psi},B\psi)+F(h)} \rangle_{X,\Lambda}$$

where g,h are test functions supported in Λ ; B is a finite rank operator on V_F , so that

$$(\overline{\psi},B\psi) = \sum_{i=1}^{N} \overline{\psi}(g_i)\psi(f_i)$$

with some (spinorial) test functions supported in Λ . $F = \varepsilon_{\mu\nu}\partial_\mu A_\nu$ is the (euclidean) field strength. We then have

Theorem 7.12 : In Higgs_2

$$\log|Z^X_{g,h;LT}| \leq cLT + a(\|g\|_1 + \|g\|_2^2) + \frac{1}{2}\|h\|_2^2 + b \tag{7.34}$$

for $X = D_M$, F or p . In QED_2

$$\log|Z^X_{B,h;LT}| \leq cLT + \frac{1}{M}\|B\|_1 + \frac{1}{2}\|h\|_2^2 + b \tag{7.34'}$$

for $X = D_M$, F or ap .

a,b,c are constants independent of L,T .

Proof : By the analogue of (7.30) for modified partition functions one sees that it is sufficient to consider X = (a)p ; for this case the "chessboard estimate" Theorem 2.2 or rather a limiting version of it says for Higgs_2

$$\log|Z^p_{g,h;LT}| \leq \int d^2x \log Z^p_{\tilde{g}_x,\tilde{h}_x;LT} \tag{7.35}$$

where $\tilde{g}_x$ ($\tilde{h}_x$) is the constant function on Λ with value g(x) (h(x)) .

The universal term $\frac{1}{2}\|h\|_2^2$ in (7.34) arises from a so-called infrared bound (see [32]) : We write

$$e^{F(h)}dm(A) = e^{\frac{1}{2}\|h\|_{2,\lambda}^2}dm(A+a(h)) \tag{7.36}$$

where a(h) obeys

$$\varepsilon_{\mu\nu}\partial_\mu a_\nu(h) = h-\lambda^2(-\Delta+\lambda^2)^{-1}h \ ; \quad \partial_\mu a_\mu(h) = 0 \tag{7.37}$$

and $$\|h\|_{2,\lambda} = (h,(1-\lambda^2\Delta)^{-1}h)$$

Therefore, introducing temporarily an ultraviolet cutoff t

$$e^{-\frac{1}{2}\|h\|^2_{2,\lambda}} Z^P_{g,h;LT} = \lim_{t\to o}\int dm_t(A) Z^P_{g;LT}(A-a(h))e^{E^t_{LT}} \tag{7.38}$$

where $Z^P_{g;LT}(A)$ is the matter partition function in the external field A. By the diamagnetic bound and its proof (Theorem 2.6 and Lemma 2.9) $Z^P_{g;LT}(A)$ is the continuum limit of functions of positive type in the holonomy operators $e^{i\phi A}$. Therefore

$$|\int dm_t(A)\ Z^P_{g;LT}(A-a(h))| \le \int Z^P_{g;LT}(A)dm_t(A) \tag{7.39}$$

and by (7.38) and the stability expansion

$$|Z^P_{g;h;LT}| \le \exp(\tfrac{1}{2}\|h\|^2_{2,\lambda})\ Z^P_{g,0,LT} \tag{7.40}$$

Note that $\|h\|^2_{2,\lambda} \le \|h\|^2_2$.

To understand where the norms $\|g\|_2$, $\|g\|_1$ in (7.34) come from one has to look at Lemma 7.7 that estimates the dependence on δm^2 . It says that for $g(x)$ large (e.g. ≥ 1)

$$|Z^P_{\tilde{g}_x;LT}(A)| \le O\ (e^{\text{const } g(x)^2})\ .$$

If we feed this into the stability expansion we obtain (cf. [29]) for $g(x)^2 \ge 1$:

$$|Z^P_{\tilde{g}_x,0,LT}| \le e^{\text{const } g(x)^2 LT} \tag{7.41}$$

where we also used Theorem 7.10 for $Z^P_{\tilde{g}_x,0;LT}$.

On the other hand it is easy to see that $\log|Z^P_{\tilde{g}_x,0;LT}|$ is a convex function of $g(x)$ and therefore for $|g(x)| \le 1$ (see [29])

$$|Z^P_{\tilde{g}_x,0;LT}| \le e^{\text{const}|g(x)|LT}\ Z^P_{0,0;LT} \tag{7.42}$$

From these two facts one obtains (7.34). For QED_2 the strategy is similar:

<u>Lemma 7.13</u> : Let $Z^{ap}_{B;LT}(A) \equiv \det_{ren}(1+K_F(A))\langle\exp(\psi,B\psi)\rangle_F(A)$ where $\langle\cdot\rangle_F(A)$ denotes the expectation with respect to the fermions in the external gauge field (assumed to be bounded and Hölder continuous). Then

$$|Z^{ap}_{B;LT}(A)| \le \exp(\tfrac{1}{M}\|B\|_1) \tag{7.43}$$

Proof : We write

$$\begin{aligned} Z^{ap}_{B;LT}(A) &= \det_{ren}(1+K_F+G_F B) \\ &= \det_{ren}(1+K_F(A))\det(1+K_F G_F' B) \qquad (7.44) \\ &= \det_{ren}(1+K_F(A))\det(1+\sum_{i=1}^{N} g_i G_F' f_i) \quad . \end{aligned}$$

The first factor is ≤ 1 , the second factor is, by the well known inequality $|\det(1+A)| \leq \exp \|A\|_1$ (see [20]), bounded by

$$\exp(\frac{1}{M}\|B\|_1) = \exp[\frac{1}{M}\sum_{i=1}^{N}\|f_i\|_2\|g_i\|_2] \quad .$$

□

Again we have to feed this into the stability expansion. As in $Higgs_2$ we get

$$e^{-\frac{1}{2}\|h\|^2_{2,\lambda}} Z^{ap}_{B,h;LT} = \lim_{t\to o}\int dm_t(A)\, Z^{ap}_{B,h;LT}(A-a(h))\, e^{E^t_{LT}}$$

By Lemma 7.13 the expression under the $\lim_{t\to o}$ is bounded by

$$\int dm_t(A)\, e^{\frac{1}{M}\|B\|_1} e^{E^t_{LT}} = e^{\frac{1}{M}\|B\|_1 + E^t_{LT}} \qquad (7.45)$$

Feeding this bound into the stability expansion produces a bound

$$|Z^{ap}_{f,g,h;LT}| \leq e^{\frac{1}{2}\|h\|^2_2 + \frac{1}{M}\|B\|_1 + b + cLT} \qquad (7.46)$$

(by also applying Theorem 7.10) which is the bound in Theorem 7.12. □

Next we want to look at the constant c of Theorem 7.10 more closely. The goal is to show that c may be chosen to be

$$c = \lim_{L,T\to\infty} \frac{1}{LT}\log Z^X_{LT} \equiv \alpha^X_\infty \leq \alpha^{(a)p}_\infty \qquad (7.47)$$

For $X = (a)p$ this follows directly from the chessboard bound and the fact that $(Z^{(a)p}_{LT})^{1/LT}$ is essentially decreasing in L and T .

For $X = F$ it can be proven by the methods of [18] which are a simple application of Schwarz's inequality with respect to the Osterwalder-Schrader inner product (and the Reeh-Schlieder theorem [47]) and produce what might be called a "free b.c. chessboard bound". So we obtain

Theorem 7.14 : In Higgs_2

$$\log|Z^X_{g,h;LT}| \leq \alpha^X_\infty(LT-1) + a(\|g\|_2^2 + \|g\|_1) + \frac{1}{2}\|h\|_2^2 + c \tag{7.48}$$

In QED_2

$$\log|Z^X_{B,h;LT}| \leq \alpha^X_\infty(LT-1) + \frac{1}{M}\|B\|_1 + \frac{1}{2}\|h\|_2^2 + c \tag{7.48'}$$

for $X = (a)p,F$, provided g, B are supported in a unit square Δ. (This means of course that g_i, f_i are supported in Δ if $B = \sum_{i=1}^N f_i \otimes g_i$).

This implies a volume independent bound for normalized expectations:

Corollary 7.15 : In Higgs_2

$$|\langle e^{:|\phi|^2:(g)+F(h)}\rangle^X_{LT}| \leq e^{\frac{1}{2}\|h\|_2^2}\, e^{a(\|g\|_2^2 + \|g\|_1)}\, e^{-\alpha^X_\infty + c} \tag{7.49}$$

In QED_2

$$|\langle e^{(\bar{\psi},B\psi)+F(h)}\rangle^X_{LT}| \leq e^{\frac{1}{2}\|h\|_2^2}\, e^{\frac{1}{M}\|B\|_1}\, e^{-\alpha^X_\infty + c} \tag{7.49'}$$

for $X = (a)p,F$; g and B supported in Δ .

Proof : This follows from Theorem 7.14 and the volume dependent bounds on the partition functions. □

By the same method one obtains

Corollary 7.16 : In Higgs_2

$$|\langle e^{:|\phi|^2:(g)+F(h)}\rangle^X_{LT}| \leq c_o e^{\frac{1}{2}\|h\|_2^2}\, e^{a(\|g\|_1 + \|g\|_2^2)}\, e^{(-\alpha^X_\infty + c)|\text{supp } g|} \tag{7.50}$$

In QED_2

$$|\langle e^{(\bar{\psi},B\psi)+F(h)}\rangle^X_{LT}| \leq c_o\, e^{\frac{1}{2}\|h\|_2^2}\, e^{\frac{1}{M}\|B\|_1}\, e^{(-\alpha^X_\infty + c)|\text{supp } B|} \tag{7.50'}$$

for $X = (a)p,F$.

To eliminate the unpleasant terms involving supports there is a very simple argument in the case of Higgs_2 : Consider the function on $\bar{\mathbb{R}}_+ = [0,\infty)$

$$F(t) \equiv \log|\langle e^{:|\phi|^2:(g)t+F(h)}\rangle^X_{LT}| \tag{7.51}$$

It is straightforward to see that F is continuous and convex on $[0,\infty)$, $F(0) \le \frac{1}{2}\|h\|_2^2 + \log c_o$,

$$\lim_{t\to\infty} \frac{1}{t} F(t) \le a(\|g\|_1 + \|g\|_2^2) \quad .$$

These properties imply

$$F(t) \le ta(\|g\|_1 + \|g\|_2^2) + \frac{1}{2}\|h\|_2^2 + \log c_o \quad . \tag{7.52}$$

If we define

$$|||g||| \equiv \Sigma \|g\chi_\Delta\|_2 \tag{7.53}$$

(cf. [29]) where Δ runs through a paving of $\mathbb{R}^2$ by unit squares, we see that we have proven

$$|\langle e^{:|\phi|^2:(g)+F(h)}\rangle^X_{LT}| \le c_o\, e^{\frac{1}{2}\|h\|_2^2 + a|||g|||^2 + a} \tag{7.54}$$

because $|||g||| \ge \|g\|_1$, $|||g|||^2 \ge \|g\|_2^2$ and $1 + |||g|||^2 \le \max\{|||g|||\ , \ |||g|||^2\}$. By an analogous argument one can also eliminate the factor $c_o e^a$.

For QED_2 a different argument has to be used. By a Cauchy estimate we can infer from (7.50') :

$$|\langle \prod_{i=1}^N \bar\psi(g_i\chi_{\Delta_i}) \prod_{j=1}^N \psi(f_j\chi_{\Delta_j})\ e^{F(h)}\rangle^X_{LT}| \le \tag{7.55}$$

$$\le c_o a^N e^{\frac{1}{2}\|h\|_2^2} \prod_{i=1}^N \|g_i\chi_{\Delta_i}\|_2 \prod_{j=1}^N \|f_j\chi_{\Delta_j}\|_2 \quad .$$

By expanding the exponential $e^{(\bar\psi,B\psi)}$ and using

$$(\bar\psi,B\psi) = \sum_{i=1}^N \bar\psi(g_i)\psi(f_i) = \sum_{i=1}^N \sum_{\Delta,\Delta'} \bar\psi(g_i\chi_\Delta)\psi(f_i\chi_{\Delta'})$$

we obtain

$$|<e^{(\bar{\psi},B\psi)+F(h)}>_{LT}^{X}| \leq c_o\, e^{\frac{1}{2}\|h\|_2^2} \times$$

$$\times \sum_{n=o}^{\infty} \frac{1}{n!} a^n \sum_{\Sigma n_{i\Delta\Delta'}=n} \frac{n!}{\prod_{i,\Delta,\Delta'} n_{i_{\Delta\Delta'}}!} \|g_i\chi_\Delta\|_2^{n_{i\Delta\Delta'}} \|f_i\chi_\Delta\|_2^{n_{i\Delta\Delta'}} =$$

$$= c_o\, e^{\frac{1}{2}\|h\|_2^2} \sum_{n=o}^{\infty} \frac{1}{n!} a^n \Big(\sum_{i,\Delta,\Delta'} \|g_i\chi_\Delta\|_2 \|f_i\chi_{\Delta'}\|_2 \Big)^n \qquad (7.56)$$

$$\leq c_o\, e^{\frac{1}{2}\|h\|_2^2} \exp\Big[a \sum_{i=1}^{N} |||g_i||| \;\; |||f_i|||\Big]$$

with $<\cdot>$ defined in (7.53).

So we have obtained finally the following volume independent bounds :

<u>Theorem 7.17</u> : In Higgs_2

$$|<e^{:|\phi|^2:(g)+F(h)}>_{LT}^{X}| \leq e^{\frac{1}{2}\|h\|_2^2 + a|||g|||^2} \qquad (7.57)$$

In QED_2

$$|<e^{(\bar{\psi},B\psi)+F(h)}>_{LT}^{X}| \leq c_o\, e^{\frac{1}{2}\|h\|_2^2} \quad \exp\Big[a \sum_{i=1}^{N} |||g_i||| \;\; |||f_i|||\Big] \qquad (7.57')$$

for $X = (a)p,F$.

<u>Remarks</u> :

1) In [29] the bound (7.57) is proven also for $X = D_M$ by similar methods plus correlation inequalities. It can also be inferred from the proofs given here by using a form of the chessboard bound valid for Dirichlet b.c. and the fact that the periodic "pressure" α_∞^P dominates all "pressures" α_∞^X with other b.c. (cf. (7.30)).

2) It is not hard to prove by the same methods for Higgs_2

$$|<e^{\phi(f)+\bar{\phi}(g)+F(h)}>_{LT}^{X}| \leq c e^{\frac{1}{2}\|h\|_2^2 + \frac{1}{m^2}\|f\|_2 \|g\|_2} \qquad (7.58)$$

and for QED_2

$$|<e^{j_\mu(a_\mu)+F(h)}>_{LT}^{X}| \leq e^{\frac{1}{2}\|h\|_2^2 + b\|a\|_2^2} \qquad (7.59)$$

where $j_\mu = :\overline{\psi}\gamma^\mu\psi:$

3) By Cauchy estimates we obtain from (7.57), (7.57')

$$|<\prod_{i=1}^{k} :|\phi|^2:(g_i) \prod_{j=1}^{\ell} F(h_j)>_{LT}^{X}| \leq c^{k+\ell}(k!\ell!)^{1/2} \prod_{i=1}^{k} |||g_i||| \prod_{j=1}^{\ell} \|h_j\|_2^2 \tag{7.60}$$

for Higgs_2 and

$$|<\prod_{i=1}^{k} (\psi(f_i)\overline{\psi}(g_i)) \prod_{j=1}^{\ell} F(h_j)>_{LT}^{X}| \leq c^{2k+\ell}(\ell!)^{1/2} \prod_{i=1}^{k} (\|f_i|||\ |||g_i|||) \prod_{j=1}^{\ell} \|h_j\|_2^2 \tag{7.60'}$$

for QED_2 .

(7.60), (7.60') give the zeroth Osterwalder-Schrader axiom [4], namely a temperedness bound and a growth condition (in k,ℓ) for the Schwinger functions. It should be remarked, however, that the reconstruction of a relativistic field theory can be based just as well on (7.57), (7.57') directly (cf. [48]).

4) We note another consequence of (7.58), (7.57') : They allow to prove uniform bounds on expectation values of "strings"

$$S(C_{xy}) \equiv \overline{\phi}(x)\, e^{i\int_{C_{xy}} A} \phi(y) \tag{7.61}$$

or

$$S(C_{xy}) \equiv \overline{\psi}(x)\, e^{i\int_{C_{xy}} A} \psi(y) \tag{7.61'}$$

where C_{xy} is a (piecewise smooth) path from x to $y \neq x$. To see this for Higgs_2 is quite straightforward; for QED_2 one has to insert the beginning of the perturbation expansion (i.e. iterated resolvent equations) for G_F' into the Matthews-Salam formulas (6.28), (6.29) :

$$G_F' = G_F + \widetilde{K}G_F + \widetilde{K}G_F\widetilde{K}G_F + (\widetilde{K}G_F)^3 + \widetilde{K}G_F\widetilde{K}G_F'(\widetilde{K}G_F)^2 \tag{7.62}$$

where

$$\widetilde{K} = (\not{\partial}+M)^{-1}\not{A} \ .$$

For the first four terms in (7.62) a bound is trivial after all we proved; for the last term it is also not hard to see that

$$\int (\tilde{K}G_F\tilde{K}G_F'(\tilde{K}G_F)^2)(x_o,y_o)dm(A)$$

is the expectation value of $\psi(f_{x_o,A})\bar{\psi}(g_{y_o,A})$ with $f_{x_o,A}, g_{y_o,A}$ in L^2 for almost every A and

$$\int \|f_{x_o,A}\|_2^2 dm(A) \ , \ \int \|g_{y_o,A}\|_2^2 dm(A) < \infty \ .$$

Graphically $f_{x_o,A}$ corresponds to 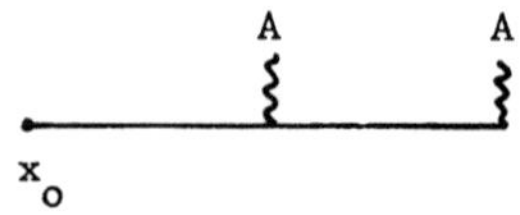

, that is

$$f_{x_o,A} = \int G_F(x_o,x')\not{A}(x')G_F(x',x)\not{A}(x)dx'$$

etc.

In the last section we will describe a reconstruction scheme based on expectations of "strings" and similar gauge invariant nonlocal objects.

c) Thermodynamic Limit; Verification of the Axioms

There are two general strategies for the thermodynamic limit : One is based on monotonicity in the volume (correlation inequalities), the other, more constructive one on the cluster expansion.

For $Higgs_2$ we are in the lucky situation of having correlation inequalities; so far, however, a cluster expansion has not been constructed. This would be very useful, but is by no means trivial, at least in the "Higgsian" regime. For QED_2 , on the other hand, we do not know any correlation inequalities, but we will sketch how in this model a cluster expansion could be constructed. This cluster expansion should also work for $Higgs_2$ in the QED-like regime $e/m \ll 1$, $\lambda/m^2 \ll 1$.

But first we have to say something about Euclidean covariance. Our states in an external gauge field were euclidean covariant in the sense that they were invariant under joint euclidean motion of the gauge field and the observables; no lattice orientation was remembered as shown in Section 6.

But our cutoff measures for the gauge field again violated euclidean covariance by singling out a time direction. So we have to show that after the limit $t \to 0$ has been taken this direction is not remembered.

It is very plausible that this should be true because

1) the violation of euclidean covariance happened only at high momenta

2) By the convergence of the stability expansion these high momenta contribute an arbitrarily small amount.

The actual proof is based on this idea :

Let $D^{t,\theta}_{\mu\nu}$ be the covariance in which the cutoff direction is rotated by an angle θ, $<\cdot>_{t,\theta}$ the corresponding cutoff expectation value. Then we have

Theorem 7.19 : For a polynomial or exponential observable P

$$\lim_{t\to o} \{Z_{t,\theta}<P>_{t,\theta} - Z_{t,o}<P>_{t,o}\} = 0 \quad . \tag{7.61}$$

Proof : By introducing an interpolating field $A(s)$ with covariance $sD^{0,t}_{\mu\nu} + (1-s)D^{\theta,t}_{\mu\nu}$ we can write the difference in (7.61) as

$$\int_o^1 ds \int dm_t(A) \int d\mu_{A(s)} KP \tag{7.62}$$

where K looks like in Subsection a). But KP is now a very small observable if t is small and (7.62) can be shown to go to zero like t^α where $\alpha > 0$. See [29] for more details. □

Now to the thermodynamic limit. First we use correlation inequalities to obtain

Theorem 7.20 : For an arbitrary sequence of rectangles $\Lambda_n \nearrow \mathbb{R}^2$, $f,g \in S(\mathbb{R}^2)$ in $Higgs_2$, $\lim_{n\to\infty} <e^{:|\phi|^2:(g)+F(h)}>^{D_M}_{\Lambda_n}$ exists and is independent of the sequence (Λ_n) .

Proof : By the correlation inequalities of Section 2.d) $<e^{-:|\phi|^2:(g)+F(f)}>^{D_M}_{\Lambda}$ is decreasing in Λ provided $g \geq 0$. By the bound (7.55) the family of entire functions on $\mathbb{C}^3$

$$\{F_n(z_1,z_2,z_3)\} \equiv \{<e^{z_1F(h)+:|\phi|^2:(z_2g_+)-:|\phi|^2:(z_3g_-)}>^{D_M}_{\Lambda_n}$$

is a normal family and, assuming g_+ , $g_- \geq 0$, h real, it converges for $z_2 \leq 0$, $z_3 \geq 0$, z_1 real. So by Vitali's theorem it converges everywhere (this argument is taken from Fröhlich [40]).

The convergence is uniform on compact sets and the limit is independent of the sequence. □

Corollary 7.21 : $\langle e^{-:|\phi|^2:(g)+F(h)}\rangle$ is euclidean invariant.

Corollary 7.22 : The infinite volume Schwinger functions

$$S_{m+n}(g_1,\dots,g_m;h_1,\dots,h_n) \equiv \langle \prod_{k=1}^{m} :|\phi|^2:(g_k) \prod_{j=1}^{n} (iF(h_j))\rangle$$

obey all the Osterwalder-Schrader axioms except possibly clustering.

Proof of Corollary 7.21 : This follows from Theorem 7.19 and the independence of the limit of the sequence (Λ_n).

□

Proof of Corollary 7.22 :

0) The zeroth axiom requires some temperedness that is implied by (7.59).

1) Symmetry is obvious.

2) Euclidean invariance is Corollary 7.21.

3) Osterwalder-Schrader positivity follows from the lattice approximation (some care is needed for the Gaussian measure; see [28]).

What we have constructed so far does not yet really correspond to Higgs_2 because of the "photon mass" λ that is still present. This can also be removed by correlation inequalities : As λ^2 decreases the transverse part of $D_{\mu\nu}$ increases; the longitudinal part is irrelevant due to the gauge invariant coupling between matter and gauge fields.

So by the correlation inequalities of Section 2.d) $\langle e^{-:|\phi|^2:(g)+F(h)}\rangle$ is increasing as $\lambda^2 \to 0$ (for $g \geq 0$) and we only need an upper bound.

This is not trivial because for $\lambda^2 \to 0$ the covariance $D_{\mu\nu}$ becomes ill-defined due to a infrared singularity. The reason why an upper bound can be obtained is that 1) by a correlation inequality one can go into a finite box and 2) there the dangerous zero mode of $D_{\mu\nu}$ is isolated and decouples due to a Ward identity (i.e. gauge invariance of the coupling). This remarkable fact is a hint of the Higgs mechanism that is supposed to generate a mass gap. For details of the slightly tricky proof we refer to [29]. We just state :

Theorem 7.23 : The limits as $\lambda^2 \to 0$ of the Schwinger functions of Corollary 7.22 exist and obey all Osterwalder-Schrader axioms except possibly clustering.

Remark : A cluster expansion that would reveal the Higgs mechanism in full by showing the existence of a mass gap requires (at least in the Higgsian regime) a detailed understanding of the role played by classical solutions (instantons) which in this case are the so-called vortices [49,50], or at least topological excitations lying "close" to these vortices. This is an important problem whose solution might shed some light also on the role of classics and topology in other gauge quantum field theories.

Let us now turn to QED_2 and sketch the cluster expansion there. This will give the Wightman axioms plus a mass gap for gauge invariant local fields. Continuum cluster expansions were pioneered by Glimm, Jaffe and Spencer [38] and applied to fermion ($Yukawa_2$) models for instance by Magnen and Sénéor [46] and by Cooper and Rosen [39]. They fit also rather nicely into the general polymer framework developed in Section 3 (cf. [51, 52]). They generally work well for theories that are weak perturbations of massive free (i.e. Gaussian) models.

In the continuum it is not possible to expand in all terms that couple different points, links etc. Instead one uses the fact the free expectations are completely characterized by Green's functions (covariances); if one replaces these by modified ones that decouple different squares (cubes, hypercubes) in a paving of $\mathbb{R}^d$, the perturbed Gaussian expectation also decouples.

This can be achieved for instance by imposing 0-Dirichlet boundary conditions on the boundaries of squares (cubes, hypercubes). If we symbolize each face of a square etc. by a link of the dual lattice it becomes apparent how to imitate the cluster expansions of Section 3. The small quantity in which the expansion proceeds is the difference between the modified and unmodified Green's functions; polymers are connected sets of links in the dual lattice or, equivalently, face-connected sets of squares (cubes...) of the original paving.

To make the activity $z(\gamma)$ of such a polymer γ small of the order $e^{-b|\gamma|}$ (b large), one needs

(1) the difference between coupled and decoupled expectations to be small, which is true if the underlying free theory has large mass, i.e. strong exponential decay,

(2) the theory to be approximately Gaussian.

The size of the grid can be adjusted to optimize convergence.

In QED_2 , (1) and (2) are fulfilled provided $e/M \ll 1$, where e is the charge of the fermion field (i.e. the gauge coupling constant called g or g_o

earlier).

It is convenient to rescale the gauge field : $A \mapsto eA$. This moves e from the free action into the determinant. We have

$$\log \det_{ren}(1+K_F(eA) = \log \det_4(1+K_F(eA)) - e^2 \, Tr_{ren} K_F(A)^2 \quad . \tag{7.63}$$

$Tr_{ren} K_F(A)^2$ can be computed (see for instance Weingarten and Challifour [26]) :

$$Tr_{ren} K_F(A)^2 = \frac{1}{\pi} \int (\delta_{\mu\nu} - \frac{k_\mu k_\nu}{k^2}) \times \overline{\hat{A}_\mu(k)} \hat{A}_\nu(k) T(k^2) d^2k \tag{7.64}$$

where (with $k = \sqrt{k^2}$)

$$T(k^2) = 1 - \frac{k}{k(4M^2+k^2)^{1/2}} \quad \text{ar th} \frac{k}{(4M^2+k^2)^{1/2}} \geq 0 \tag{7.65}$$

Note that $T(k^2) = \frac{k^2}{4M^2} = O(k^4)$ for k^2 small and

$$T(k^2) = 1 - \frac{2M^2}{k^2} \log \frac{k^2}{M^2} + O(k^{-4}) \text{ , for } k^2 \text{ large.}$$

We make up a new Gaussian measure $dm(F)$ for the field strength $F = \varepsilon_{\mu\nu}\partial_\nu A_\mu$ by absorbing this quadratic term (this is not essential but convenient) : $dm(F)$ has mean zero and covariance $C(x)$ given by

$$\hat{C}(k) = \frac{k^2}{k^2+T(k^2) \frac{e^2}{\pi}} \equiv G(k^2) \tag{7.66}$$

(to arrive at this we set the "photon" mass λ equal to zero since it is of no use here).

It is easy to see that $G(k^2) \equiv \hat{C}(k^2)$ is analytic for $|\text{Im } k^2| < 4M^2$ and therefore by the Paley-Wiener theorem [33] $C(x)$ will decay essentially like $e^{-2M|x|}$. To make the analytic structure more transparent, we write with $z = \frac{k^2}{4M^2}$

$$T(k^2) = 1 - \frac{1}{z} \sqrt{\frac{z}{1+z}} \text{ ar th} \sqrt{\frac{z}{1+z}} = 1 + \frac{1}{2z} \sqrt{\frac{z}{1+z}} \log \frac{-\frac{z}{\sqrt{1+z}} + 1}{\frac{z}{\sqrt{1+z}} + 1} \tag{7.67}$$

It is now easy to see that $G(k^2)$ is a Herglotz function (i.e. has positive imaginary part in the upper half plane) and we have a Lehmann-Källén representation with a finite positive measure $d\rho(\mu^2)$:

$$G(k^2) = 1 - \int_{4M^2}^{\infty} d\rho(\mu^2) \frac{1}{\mu^2+k^2} \tag{7.68}$$

$d\rho$ obeys the following sum rules

$$\int_{4M^2}^{\infty} d\rho(\mu^2) = \frac{e^2}{\pi}$$

$$\int_{4M^2}^{\infty} d\rho(\mu^2) \frac{1}{\mu^2} = (1 + \frac{e^2}{4\pi M^2})^{-1}$$

which are obtained by looking at the behavior of (7.68) at

$$k^2 = \infty \qquad \text{and} \qquad k^2 = 0 \ .$$

Note that as in the massless Schwinger model, cf. eq. (5.20), the Gaussian field F with this covariance is "Nelson-Symanzik positive" (i.e. comes from a positive measure) but Osterwalder-Schrader negative when considered as an ordinary scalar field! This has to be so, of course, because the physical field is $E = iF$ as remarked earlier.

The full measure for F in QED_2 is (in a region Λ)

$$d\mu_\Lambda(F) = \frac{1}{Z_\Lambda} \det{}_4(1+eK_{F,\Lambda}(A))dm_\Lambda(F) \tag{7.69}$$

where $K_{F,\Lambda}(A)$ is defined by introducing 0-Dirichlet boundary conditions on $\partial\Lambda$ in G_F and dm_Λ has the covariance

$$C_\Lambda^D(x,y) \equiv \chi_\Lambda(x)\delta(x-y) - \int_{4M^2}^{\infty} d\rho(\mu^2)(-\Delta_\Lambda^D+\mu^2)^{-1}(x,y) \tag{7.70}$$

where $(-\Delta_\Lambda^D+\mu^2)^{-1}$ has 0-Dirichlet data on $\partial\Lambda$.

Actually some arguments are needed to show that the measure $d\mu_\Lambda$ defined in (7.69) is really the same as the measure $d\mu_\Lambda$ defined earlier; this is not particularly hard to see but we do not want to spend our time with this rather uninteresting question here.

According to our scheme we should really carry an ultraviolet cutoff t along, but it is easy to see (and essentially shown in [26]) that $\det_4$ can be unambigously defined as a dm_Λ-measurable function by taking the limit $t \to 0$. Our stability expansion actually shows that it is L^1 . Furthermore it is gauge invariance that allows to consider $\det_4(1+eK_{F,\Lambda}(A))$ as a function of F .

As discussed earlier, it is possible to consider gauge invariant "string-like" objects involving fermions which are represented here as

$$G'_F(C_{xy};F) \equiv e^{i\int_{C_{xy}} eA_\mu dx_\mu} G'_F(x,y;A)$$

The electromagnetic current can be obtained as in the massless Schwinger model by introducing an external vector potential a and taking functional derivatives, So for instance

$$<e^{iej_\mu(a_\mu)}>_\Lambda = \frac{1}{Z_\Lambda} \int \det_4(1+K_{F,\Lambda}(A+a))dm_\Lambda(F+f) \tag{7.72}$$

where $f = \varepsilon_{\mu\nu}\partial_\mu a_\nu$.

In (7.71) and (7.72) 0-Dirichlet b.c. on $\partial\Lambda$ are to be used in G_F , G'_F , and both (7.71) and the integrand in (7.72) can be understood as functions of F .

So from now on we will be content with discussing expectations of observables that are functions of F .

To get a cluster expansion we now introduce a paving of Λ by unit squares Δ and we introduce decoupled Green's functions. To avoid any trouble with gauge invariance - which could spoil the whole game here - we use 0-Dirichlet b.c. on the boundaries of our unit squares. Neumann b.c. would be more directly analogous to the lattice expansion but they are not as well analyzed as are Dirichlet b.c., in particular for fermions (see Cooper and Rosen [39]).

To set up the expansion we denote by $C^{\mathcal{B}}$, $G_F^{\mathcal{B}}$ the Green's functions with Dirichlet b.c. on all links <u>not</u> in the set $\mathcal{B}$ of links. Furthermore we define

$$C(\underline{s}) \equiv \sum_{\mathcal{B}\subset\mathcal{B}\Lambda} \prod_{b\in\mathcal{B}} s_b \prod_{b\notin\mathcal{B}} (1-s_b)C^{\mathcal{B}} \tag{7.73}$$

$$G_F(\underline{s}) \equiv \sum_{\mathcal{B}\subset\mathcal{B}_\Lambda} \prod_{b\in\mathcal{B}} s_b \prod_{b\notin\mathcal{B}} (1-s_b)G_F^{\mathcal{B}} \tag{7.73'}$$

where $\mathcal{B}_\Lambda$ is the set of all links in Λ .

Note that $C(\underline{0})$, $G_F(\underline{0})$ decouple all squares, whereas $C(\underline{1})$, $G_F(\underline{1})$ are the

Green's functions with Dirichlet b.c. only on $\partial\Lambda$.

The following formula is an obvious identity (the fundamental theorem of calculus) :

$$C(\underline{1}) = \left(\sum_{\Gamma} \prod_{\gamma \in \Gamma} \prod_{b \in \gamma} \int_0^1 ds_b \frac{\partial}{\partial s_b} \right) C(\underline{s}_{\Gamma}) \tag{7.74}$$

where the sum is over all sets of disjoint polymers (connected sets of links in the dual lattice) in Λ and $\underline{s}_{\Gamma}$ replaces s_b by 0 for all b not in any $\gamma \in \Gamma$. There is of course a similar formula for G_F .

If Γ is a disjoint union of $\gamma_1, \dots, \gamma_n$, the corresponding partition functions factorize :

$$Z_{\Gamma} = \prod_{i=1}^{n} Z_{\gamma_i} \tag{7.75}$$

For "modified partition functions" (unnormalized expectations) the situation is analogous except that anything hitting the support of the observable is lumped into one polymer (observables involving G_F , like (7.71) have to be decoupled too).

The activities are now defined as

$$z(\gamma) \equiv \left(\prod_{b \in \gamma} \int_0^1 ds_b \frac{\partial}{\partial s_b} \right) Z_{\gamma}(\underline{s}_{\gamma}) \tag{7.76}$$

and similarly modified activities

$$z_P(\gamma) \equiv \left(\prod_{b \in \gamma} \int_0^1 ds_b \frac{\partial}{\partial s_b} \right) Z_{\gamma}(\underline{s}_{\gamma}) \langle P \rangle_{\underline{s}_{\gamma}} \tag{7.77}$$

As in Section 3 we now obtain

$$\langle P \rangle_{\Lambda} = \sum_X{}' z_P^X \frac{a(X)}{X!} \tag{7.78}$$

where the sum Σ' is over all multi-indices (clusters) connected to P and linear in it.

As we saw in Section 3, for uniform (in Λ) and absolute convergence of (7.78) one only has to check the bound

$$|z_P(\gamma)| \leq c e^{-b|\gamma|} \tag{7.79}$$

with sufficiently large b .

To prove (7.79) one uses integration by parts as in the stability expansion to evaluate

$$\frac{\partial}{\partial s_b} Z(\underline{s}_\gamma) \langle P \rangle_{\underline{s}_\gamma} \equiv \frac{\partial}{\partial s_b} \int d\mu_{\underline{s}_\gamma}(F) P \tag{7.80}$$

This gives

$$\frac{\partial}{\partial s_b} Z(\underline{s}_\gamma) \langle P \rangle_{\underline{s}_\gamma} = \int d\mu_{\underline{s}_\gamma}(F) K_b P$$

where in the pictorial language used before

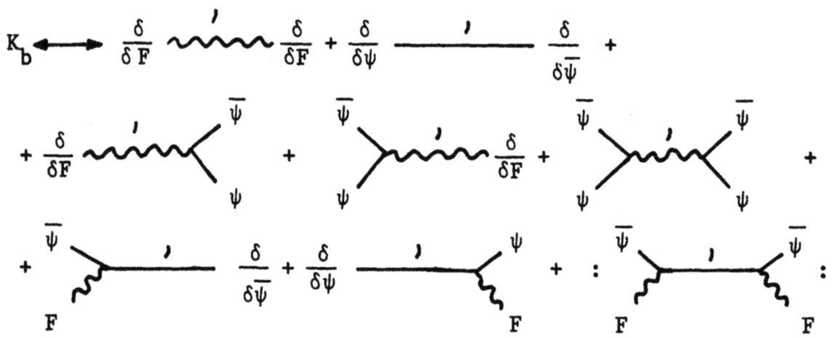

(7.81)

Here  stands for $\frac{\partial}{\partial s_b} C(\underline{s}_\gamma)$ and  for $\frac{\partial}{\partial s_b} G_F$.

The Wick ordering in the last term refers both to the fermions and the "photons"; the fermions are to be Wick ordered with respect to the free Green's function and for the "photon" field F one may Wick order with respect to its covariance C or the white noise covariance. This Wick ordering arises from the fact that the measure involves $\det_4$ and it eliminates all dangerous (divergent or conditionally convergent) graphs. The fermions in the vertex  may be considered to be Wick ordered (formally their Wick subtraction vanishes).

One important remark has to be made about the vertex 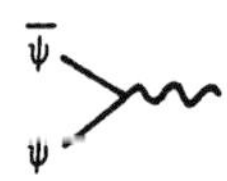 F . As remarked it stands for :

$$:\bar{\psi}\gamma_\mu\psi:\ (\varepsilon_{\mu\nu}\Delta^{-1}\partial_\nu F)$$

and this looks dangerously long range. The reason why it is alright is gauge invariance : $:\bar{\psi}\gamma_\mu\psi:$ will always contract to some other fermions, producing a convergent Feynman graph (all possibly dangerous ones are absorbed in the covariance of F) like

$\equiv f_\mu$

obeying

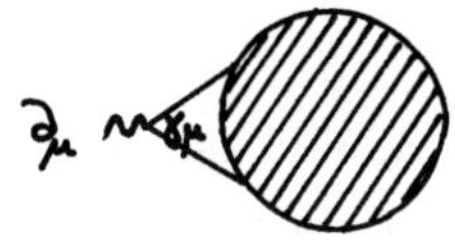

$= \partial_\mu f_\mu = 0$

(7.82)

But this means (since we have no harmonics in the Hodge decomposition due to our boundary conditions)

$$f_\mu = \varepsilon_{\mu\nu}\partial_\nu g$$

so the suspicious looking Δ^{-1} in the vertex is always eaten up. Of course its presence is essential for obtaining good power counting, i.e. , finite graphs.

This little consideration shows that gauge invariance in the form of Ward identities is crucial for the mass generation in this model.

The estimation of

$$z_P(\gamma) = \int d\mu_{\underline{s}_\gamma} \left(\prod_{b\in\gamma} K_b \right) P \tag{7.83}$$

proceeds now in a fashion that is well known in constructive field theory and up to a certain point parallels the stability expansion . The derivatives present in K_b make (7.83) small in two ways :

$$\prod_{b\in\mathcal{B}} \frac{\partial}{\partial s_b} C(\underline{s}_\gamma)(x,y) \quad \text{and} \quad \prod_{b\in\mathcal{B}} \frac{\partial}{\partial s_b} G_F(\underline{s}_\gamma)(x,y)$$

become exponentially small with $|\mathcal{B}|$ and they decay exponentially with the distance of the arguments x or y from $\mathcal{B}$.

For the necessary estimates on the differentiated Green's functions that are not at all trivial, as well as for the remaining combinatorical estimates we refer to the beautiful article [39].

A bound of the form (7.79) emerges, provided e is small and M is large. By scale covariance it is clear that only the ratio $\frac{e}{M}$ is relevant (one has to vary the size of the squares in the paving to see this). b will be of the order of $2M$ because our observables are even in the fermion fields.

I think I have made clear how the following could be proven :

Quasy-Theorem 7.24 : In QED_2 for e/M small the fields iF , j_μ obey all the Osterwalder-Schrader axioms including clustering. There is a mass gap of the order $2M$.

Remarks : (1) Notice that the expansion does not work near the trivial Schwinger limit $M \to 0$. It would be interesting to find out what is happening there. It seems that a different mass generation mechanism takes over. Unfortunately perturbation theory in M is rather singular (as opposed to the situation analyzed in [13]) as can be seen already by analyzing $\hat{C}$: it goes like $M^2 \log M$.

(2) θ-states can be defined by adding the customary term $i\theta \int F$ to the Lagrangian. The dynamics will be θ-dependent (cf. [13]).

(3) Fractionally charged Wilson-loops show area decay (as can be shown by the cluster expansion), whereas for the massless Schwinger model they show perimeter decay.

(4) The particle structure is not fully clear. Should the branch point present in $G(k^2)$ persist for the full two-point function this would suggest that there is no confinement but that screened fermions exist as particles. Obviously these questions require more study.

(5) A cluster expansion along these lines also seems to work in $Higgs_2$, but only in the "non-Higgsian" region where e^2/m^2 is small (m^2 is a positive mass term for the Higgs field) and the self-interaction is weak. This leaves again the question open what is going on in the Higgs region. The lattice theory does not give any hint of a phase transition between the two regions.

This concludes our study of the construction of continuum gauge theories. As became apparent the successes so far are severely limited : The program could be pushed through to the end (i.e. the Wightman/Osterwalder-Schrader axioms) only for two dimensions and abelian gauge group. We tried to make clear, however, that many of the intermediate results are of a more general nature and that work is in progress to get at least beyond the flat two-dimensional world [9,16].

If one wants to construct a nonabelian model like QCD in $d = 2$ or 3 (dare I mention $d = 4$?), there is, however, another problem that cannot just be overcome by sufficient skill in proving estimates : It is the lack of a sufficient number of fields for which the Wightman axioms can be expected to hold. It seems appropriate to look for a more general framework that is more adapted to the special features of gauge theories. This is the subject of the last section.

8. A FRAMEWORK FOR NONLOCAL GAUGE INVARIANT OBJECTS.

Even though it is possible to show that gauge invariant local fields obey the Wightman axioms in some cases, as we have seen, it seems that these axioms do not provide the most natural framework for theories like QCD . We saw already in the discussion of the scaling limit that is is best discussed for expectations of objects like Wilson loops. It should also be stressed that such nonlocal gauge invariant objects are much closer to the expected particle content of confining theories. So the gauge invariant string

$$S(C_{xy}) \equiv \bar{\psi}(x) P\, e^{i\int_{C_{xy}} A} \psi(y) \tag{8.1}$$

(represented pictorially by x  y)

seems a natural candidate for creating mesons; of course the Wilson loops would be expected to create primarily glueballs (if there are any) and maybe also mesons because states $S(C_{xy})\Omega$ and $W(C)\Omega$ will not be orthogonal in general.

There is also a natural candidate for creating baryons in QCD (of course with color group SU(3)) :

$$B(C_{x_1y}, C_{x_2y}, C_{x_3y}) \equiv \psi_a(x_1)\psi_b(x_2)\psi_c(x_3) \times$$

$$(P\, e^{i\int_{C_{x_1y}} A})_{aa'}\, (P\, e^{i\int_{C_{x_2y}} A)}{}_{bb'}\, (P\, e^{i\int_{C_{x_3y}} A})_{cc'} \quad \varepsilon_{a'b'c'} \tag{8.2}$$

where the indices refer to color space. Pictorially B may be represented as a "star"

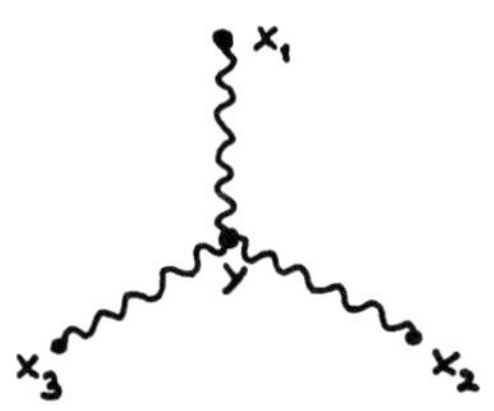

If we assume that we manage to construct euclidean continuum expectations of such things, be it as a scaling limit or by solving the Schwinger-Dyson equations as advocated by many people or in any other way [54] we have to face the question what this means. Does this determine a quantum field theory of some sort ? Can we describe scattering of particles this way ?

It is this kind of question that is addressed in [55] and we want to give some ideas here how these questions are answered in essence positively, provided one is willing to make suitable assumptions - that seem at least plausible - about the expectation values.

It should be stressed that this is a general framework that may also be suited for other "field theories" of non-local objects such as "dual string theory" [56,57] or some "bag models" [58].

To keep the notation simple, I will formulate everything explicitly only for Wilson loops. Sometimes I will remark on changes that are necessary when dealing with "strings" or "stars", but generally everything goes through for these without change.

a) Assumptions.

These assumptions are partly of a somewhat complicated technical nature and one should keep an open mind about them and not consider them sacred (this is why we do not call them axioms). The point is that there are assumptions that are not unreasonable and that allow to carry out the constructions we are presenting. More general assumptions would probably still allow for similar results, but at much higher cost in person-hours.

Wilson loops in principle carry a group representation label τ which I will suppress. Loops are always assumed to be at least piecewise smooth. We call

$$S_n(C_1,\ldots,C_n) \equiv \langle \prod_{i=1}^{n} W(C_i)\rangle \tag{8.3}$$

the n-loop Schwinger function.

The assumptions about S_n that roughly parallel the Osterwalder-Schrader axioms [4] are

(S0) Technical assumptions

(S1) Symmetry

(S2) Euclidean Invariance

(S3) Osterwalder-Schrader positivity

(S4) Clustering

Remark : If "baryon operators" ("stars") are present, (S1) has of course to be modified according to their Fermi statistics.

We now want to give the assumptions in more detail and try to explain why we consider them plausible :

(S0) :

a) $S_o = 1$, S_n is defined for piecewise smooth loops.

b) S_n depends continuously on $C_1,\dots,C_n$ in a suitable topology for loops. One such topology for $d = 4$ would be the following :

Define a distance function on parametrized loops $x(s)$ by

$$d(x(\cdot),x'(\cdot)) \equiv \|x-x'\|_2 + \|\dot{x}-\dot{x}'\|_2 + \|\ddot{x}-\ddot{x}'\|_2 \tag{8.4}$$

where the dot denotes derivative with respect to s and the last term has to be interpreted appropriately if the loops have corners (if $\ddot{x}-\ddot{x}'$ contains a δ-function, define $d(x(\cdot),x'(\cdot))$ to be ∞).

If we now identify parametrized loops $x(\cdot) \sim x'(\cdot)$ whenever the corresponding point sets coincide, we see that the sets $\{x(\cdot)|x(\cdot) \sim x_o(\cdot)\}$ are closed and the quotient topology makes sense.

Remarks : (1) The reason for choosing this topology is the following : In the free electromagnetic field theory in $d = 4$ ($QED_{4,f}$) the loops can be studied explicitly; formally

$$\log\langle W(C)\rangle = - \int_C dx^\mu \int_C dy^\nu \, D_{\mu\nu}(x-y) \tag{8.5}$$

To make this well defined it has to be renormalized by subtracting formally an infinite constant (superficially proportional to $|C|$) . But a little inspection shows that each corner makes an extra renormalization necessary, depending on the angle at the corner. The renormalized version of (8.5) cannot be continuous in a topology that allows smooth (C^1) loops to converge to loops with corners. (8.4) prevents that and it is not hard to see that (8.5) (renormalized) does in fact have the required continuity (cf. Polyakov [54], Fröhlich [55]).

(2) If one chose a much stronger topology and for instance required S_n only to be continuous in the parameters of rigid euclidean motions of the loops, the resulting physical Hilbert space (see below) could become intolerably large (i.e. nonseparable). Note that the space of loops is separable with our topology.

(3) Because of asymptotic freedom it is plausible that these considerations for $QED_{4,f}$ are valid in QCD_4 as well.

c) Let $C_i^{a_i}$ be the translate of C_i by the amount $a_i \in \mathbb{R}^d$ $(i = 1,\dots,n)$; let $d_n(C_1^{a_1},\dots,C_n^{a_n})$ be the minimal euclidean distance in $\mathbb{R}^d$ between any two of the loops. Then there are constants K_n , c_n , p such that

$$|S_n(C_1^{a_1},\dots,C_n^{a_n})| \leq K_n \exp(c_n d_n^{-p}) \tag{8.6}$$

K_n and c_n may depend on the loops chosen (but of course not on $a_1,\dots,a_n$) ; p may not.

Remark : By looking at $QED_{d,f}$ and appealing to asymptotic freedom one guesses : $p = 0$ in $d = 2$; p arbitrarily small in $d = 3$; $p = 1$ in $d = 4$.

d) Select a direction called "time"; define a temporal distance d_t between any two loops C, C' by

$$d_t(C,C') \equiv \inf\{|t-t'| \mid \exists\ (t,\vec{x}) \in C \text{ and } (t',\vec{x}') \in C'\}$$

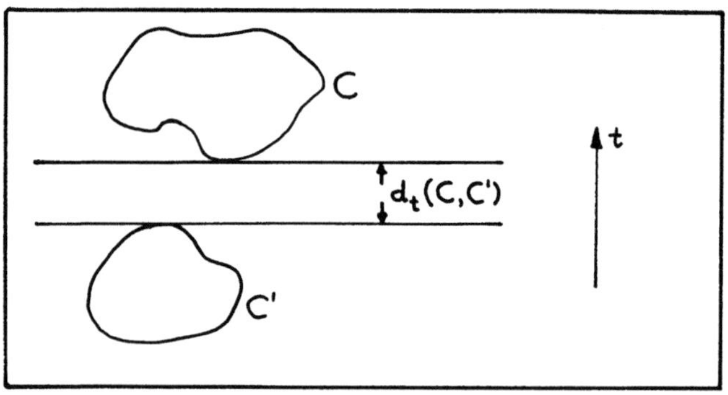

Then there are "constants" $K_{C,\varepsilon}$ (i.e. functions of C and $\varepsilon > 0$) such that

$$|S_n(C_1,\dots,C_n)| \leq K_{C_1,\varepsilon}\cdots K_{C_n,\varepsilon} \tag{8.7}$$

whenever $d_t(C_i,C_k) \geq \varepsilon$ for all $i,k = 1,\dots,n$.

Remarks : (1) Of course $K_{C,\varepsilon}$ will blow up as $\varepsilon \to 0$ in a way indicated under c).

(2) (8.7) expresses some kind of "clustering"; the point is really that $K_{C,\varepsilon}$ does not depend on n .

(3) (8.7) is of course true on the lattice; if the continuum Wilson loops only require individual multiplicative renormalization as expected, (8.7) should be true.

(4) (8.7) is true for $QED_{d,f}$. If one believes in a mass generation mechanism in QCD, it should be true there a fortiori.

(5) In a scalar field theory (8.7) would correspond to a very weak form of a ϕ - bound. It would say that

$$e^{-\varepsilon H}\phi(x)\; e^{-\varepsilon H}$$

is a bounded operator. In all existing models much stronger bounds are true.

(6) (8.7) is true (trivially) for the two-dimensional models we have discussed. There actually $K_{C,\varepsilon} = \chi_\tau(\mathbb{1})$.

(S1), (S2) need no comment except for the one made already about "stars" $B(C)$ obeying Fermi statistics.

(S3) is well known but we want to give a precise formulation : Let $V_{\neq}$ be the vector-subspace of the polynomial algebra $\mathbb{C}[\{W(C)\}]$ spanned by monomials containing only nonintersecting loops. Then the set of Schwinger-functions $\{S_n\}$ induces a linear functional S on $V_{\neq}$.

Let $V_{\pm}$ be the subspace of $V_{\neq}$ spanned by monomials containing only loops lying in $\mathbb{R}^d_{\pm} \equiv \{(t,\vec{x}) \in \mathbb{R}^d | t \gtrless 0\}$ and let θ be the antilinear map from V_+ to V_- reflecting all loops in $t = 0$ and taking complex conjugates of the coefficients. Then (S3) says

$$S(A\theta A) \geq 0 \qquad \text{for any } A \in V_+ . \tag{8.8}$$

(S4) says simply

$$\lim_{a\to\infty} S(AB^a) = S(A)S(B) \tag{8.9}$$

where B^a denotes the translate of B by a vector $a \in \mathbb{R}^d$.

b) Reconstruction of a Relativistic Quantum Mechanics.

This subsection is of a very general nature. It is really a group theoretic reconstruction theorem and only depends on (S2), (S3), (S4) ((S4) actually could even be relaxed). It can easily be generalized to other systems described by Euclidean expectations of nonlocal objects (such as bag models, for instance). We use the following

Def. 8.1 : A relativistic Quantum Mechanics consists of

(1) A separable Hilbert space $\mathcal{H}$

(2) A distinguished vector $\Omega \in \mathcal{H}$, called vacuum

(3) A continuous unitary representation of $\tilde{\mathcal{P}}_{+}^{\uparrow}$, the universal covering group of the proper orthochronous Poincaré group, obeying the "spectral condition" (spectrum of the generators of translations P_{μ} in $\overline{V}_{+}$, the closed forward light cone), with Ω as the only invariant vector.

To verify this definition in our framework we first have to give $\mathcal{H}$: Let

$$\mathcal{N} \equiv \{A \in \mathcal{V}_{+} | S(A\theta A) = 0\} \quad .$$

Then

$$\mathcal{H} \equiv \overline{\mathcal{V}_{+}/\mathcal{N}} \tag{8.9}$$

is a separable Hilbert space.

This is of course standard. It is also standard and straightforward (cf. [4]), to see by Schwarz's inequality and the bound (8.6) that translations in positive time direction induce a semigroup of positive contractions

$$\{T_t\}_{t\geq o} \quad \text{on } \mathcal{H} \text{ and that } T_t = e^{-tH} \text{ where } H \geq 0 \tag{8.10}$$

Furthermore spatial euclidean motions (a,R) (i.e. the subgroup of the euclidean motions of $\mathbb{R}^d$ that maps the hyperplane $t = 0$ into itself) are represented unitarily and continuously (in the strong topology) on $\mathcal{H}$.

So the question is about the boosts.

Consider the following subspace $\mathcal{V}_{\alpha_0,\varepsilon}$ of $\mathcal{V}_+$:

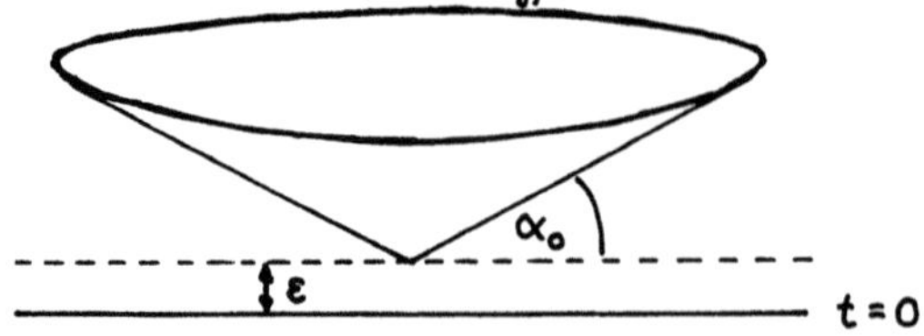

$V_{\alpha_o,\varepsilon}$ is the subspace generated by monomials containing only loops in the cone of opening angle $\pi-2\alpha_o$, apex at $x_o = (\varepsilon,\vec{0})$, which is rotationally symmetric about the time axis (see figure). $V_{\alpha_o,\varepsilon}$ corresponds to a subspace $H_{\alpha_o,\varepsilon}$ of H, and the fact that $\{e^{-tH}\}$ extends to a holomorphic semigroup in the right half plane shows by a standard argument related to the Reeh-Schlieder theorem [47] that $H_{\alpha_o,\varepsilon}$ is dense in H.

Now there is a whole neighborhood U of the identity in E_d, the euclidean group on $\mathbb{R}^d$, that acts naturally on $H_{\alpha_o,\varepsilon}$ and is represented by linear (partly unbounded) operators. Let us focus on a one-parameter subgroup $H \subset E_d$ of rotations in a 2-plane containing the time axis. We call H a group of <u>imaginary boosts</u>.

Also let $H_U \equiv H \cap U$.

<u>Lemma 8.2</u> : H_U is represented on H by a strongly continuous local group of symmetric (unbounded) operators $\{\overset{o}{P}_\alpha\}_{|\alpha|<\alpha_o}$ defined on $H_{\alpha_o,\varepsilon}$.

<u>Proof</u> : By the structure of E_d and the invariance of S we have for $A,B \in V_{\alpha_o,\varepsilon}$:

$$S(A_\alpha \theta B) = S(A(\theta B)_{-\alpha}) = S(A\theta B_\alpha) \tag{8.11}$$

where A_α denotes the image of A under a rotation by an angle α $(|\alpha| < \alpha_o)$.

From 8.11 we see that for $A \in N \cap V_{\alpha_o,\varepsilon}$, also $A_\alpha \in N$: if $S(A\theta B) = 0$ for all $B \in V_{\alpha_o,\varepsilon}$, then

$$S(A_\alpha \theta B) = S(A\theta B_\alpha) = 0$$

for all $B \in V_{\alpha_o,\varepsilon}$.

So the operators $\overset{o}{P}_\alpha$ on $H_{\alpha_o,\varepsilon}$ corresponding to H_U are well defined and by (8.11) they are symmetric.

Strong continuity follows immediately from the group property and symmetry of $\overset{o}{P}_\alpha$ together with the continuity (SO.b). □

At this point a remarkable theorem comes in handy that is due to Fröhlich [59] in the form used here; actually it can also be deduced from an older result of Glaser [60], and a variant of it has also been proven by Klein and Landau [59].

Theorem 8.3 :

Consider a semigroup $\{\overset{o}{P}_t\}_{t\geq o}$ of (possibly unbounded) linear operators $\overset{o}{P}_t$ on a separable Hilbert space $\mathcal{H}$ with the property : There is a dense linear subspace $\mathcal{D}$ of $\mathcal{H}$ such that

(1) For each $\varphi \in \mathcal{D}$ there is a $\varepsilon(\varphi) > 0$ such that φ is in the domain of $\overset{o}{P}_t$ for all $t \in [0,\varepsilon(\varphi))$ and $s - \lim\limits_{t\to o} \overset{o}{P}_t\varphi = \varphi$.

(2) If $\varphi \in \mathcal{D}$ and $s,t,s+t \in [0,\varphi(\varepsilon))$ then $\overset{o}{P}_s\varphi$ is in the domain of $\overset{o}{P}_t$ and

$$\overset{o}{P}_t\overset{o}{P}_s\varphi = \overset{o}{P}_{s+t}\varphi \quad .$$

(3) $\overset{o}{P}_t$ is symmetric for each $t \geq 0$.

Then these operators $\overset{o}{P}_t$ have a unique selfadjoint extension P_t and $\{P_t\}_{t\geq o}$ is a semigroup of selfadjoint operators on $\mathcal{H}$.

An immediate consequence is

Cor. 8.4 : There is a unique group of selfadjoint operators

$$\{P_\alpha\}_{\alpha\in\mathbb{R}} \qquad \text{such that } P_\alpha \text{ for } |\alpha| < \alpha_o$$

is an extension of $\overset{o}{P}_\alpha$.

Remark : The group $\{P_\alpha\}_{\alpha\in\mathbb{R}}$ does not represent the subgroup H but its universal covering group H (which is isomorphic to $\mathbb{R}$).

Now we can of course analytically continue the group $\{P_\alpha\}$. We write

$$P_\alpha \equiv e^{\alpha B} \tag{8.12}$$

(actually Fröhlich's proof goes by the construction of B) and call B the generator of a boost in the chosen direction. Unitary boosts are then defined as $e^{i\alpha B}$ and to make the direction dependence clear, we write

$$e^{i\vec{\alpha}\,\vec{B}} \quad .$$

Of course we also have the unitary time translations e^{itH} , so that we have candidates for unitary representatives of each element $(a,\Lambda) \in P_+^\uparrow$ by taking appropriate products of translations, rotations and boosts.

What is not yet clear at this point is that we actually have a representation. It is easy to see that the generators form a self-adjoint representation of the Poincaré Lie algebra, but it is well known that additional conditions have to be met so that the exponentials represent the group (cf. [61,62]). These conditions seem to be hard to check here. So we follow a different strategy : We know already that we have a "mixed local representation" of E_d on $\mathcal{H}$, i.e. a full neighborhood of $\mathbb{1} \in E_d$ is represented by partly unbounded operators on $\mathcal{H}$ with domain $\mathcal{V}_{\alpha_0,\varepsilon}$. So the strategy is to analytically continue the group multiplication law to obtain a unitary representation of $\mathcal{P}_+^\uparrow$. This requires some care because we cannot arbitrarily continue in all group parameters.

We start with the homogeneous part SO(d) of E_d .

Let g = O(d) be the real Lie-algebra of SO(d) , h = o(d-1) the subalgebra corresponding to the spatial rotations, and m the vector subspace of g spanned by the generators of the remaining rotations (called imaginary boosts above). Obviously

$$[h,h] \subset h \ , \ [h,m] \subset m \ , \ [m,m] \subset h \tag{8.13}$$

<u>Remark</u> : (8.13) means that the pair (g,h,σ) where σ is -1 on m , +1 on h and linear is a so-called symmetric Lie algebra (cf. [63]).

The structure (8.13) is necessary for any analytic continuation of representations.

If we replace m by im we get on account of (8.13) another real Lie algebra, called g^* . In our case it is o(d-1,1) .

We need a few simple lemmas.

<u>Prop. 8.5</u> : Each $g \in U \subset SO(d) \equiv G$ where U is a sufficiently small neighborhood of $\mathbb{1} \in SO(d)$ may be (uniquely) written in the form

$$g = \exp(tY)\,\exp(sX) \tag{8.14}$$

with $X \in h$, $Y \in m$.

<u>Note</u> : We use the mathematician's convention for the Lie algebra here (no i's).

<u>Proof</u> : If U is small enough, g belongs to precisely one coset (right orbit) gH where H = SO(d-1) .

G/H is a principal fiber bundle (see appendix) and carries a canonical connection [63]; its geodesics are given by $\{e^{\tau Y}H\}_{|\tau|\leq\varepsilon}$ ($Y \in m$). If U is small enough there will be precisely one geodesic from $0 = \mathbb{1}H$ to gH and $gH = e^{tY}H$ for exactly one t (if we require that $e^{t'Y} \in U$ for $0 \leq t' \leq t$). For U small enough $e^{-tY}g \in H$ may be written (uniquely) as e^{sX}.

□

To keep our notation simple we will now blur the distinction between Lie algebra elements and the operators on $\mathcal{H}$ representing them.

Prop. 8.6. : $\mathcal{D} \equiv \mathcal{H}_{\alpha_o,\varepsilon}$ consists of analytic vectors for m .

Proof : This follows from the fact that each $\psi \in \mathcal{D}$ is in the domain of $e^{\pm tY}$ for $Y \in m$ and t small enough.

□

Prop. 8.7 : There is a dense subspace $\mathcal{D}'$ of vectors analytic for all $X \in g$.

Proof : Let $\psi \in \mathcal{D}$, τ an irreducible unitary representation of H (finite dimensional since H is compact!) and P_τ the projection in $\mathcal{H}$ onto the subspace belonging to that representation. We claim that $P_\tau\psi$ is still an analytic vector for all $Y \in m$: P_τ can be written down explicitly (cf. Section 2.b)) :

$$P_\tau\psi = d_\tau \int_H dh\chi_\tau(h^{-1})h\psi \qquad (8.15)$$

where dh is normalized Haar measure on H .

Therefore

$$\sum_{k=o}^{\infty} \frac{t^k}{k!} \|Y^k P_\tau\psi\| \leq \chi_\tau(\mathbb{1})^2 \sum_{k=o}^{\infty} \frac{t^k}{k!} \|Y^k h\psi\| =$$

$$= \chi_\tau(\mathbb{1})^2 \sum_{k=o}^{\infty} \frac{t^k}{k!} \|((\mathrm{ad}\ h)(Y))^k\psi\|$$

which is finite for small enough t since ψ is an analytic vector for $\mathrm{ad}\ h(Y)$. $P_\tau\psi$ is obviously entire analytic for h .

□

Now we can start to analytically continue.

Let $\{Y_1,\ldots,Y_k\}$ be a basis of m , $\{X_1,\ldots,X_\ell\}$ a basis of h . Then there is a polydisc $P \subset \mathbb{C}^k$ centered at 0 , such that the maps

$$\underline{f} : P \to \mathbb{C}^k$$

$$\underline{g} : P \to \mathbb{C}^\ell$$

expressing the (abstract) group multiplication law

$$e^{z_1 Y_1} \dots e^{z_k Y_k} = e^{\sum_{i=1}^{k} f_i Y_i} e^{\sum_{i=1}^{\ell} g_i X_i} \tag{8.16}$$

are holomorphic.

For purely imaginary $z_1,\dots,z_k$, $\underline{f}$ will take purely imaginary values and $\underline{g}$ real values, so that (8.16) also expresses the multiplication law of G^* (the simply connected group generated by $\mathfrak{g}^*$) .

Now we regard $\{X_i\}_{i=1}^{\ell}$, $\{Y_i\}_{i=1}^{k}$ again as linear operators on $\mathcal{H}$ and pick $\varphi,\psi \in \mathcal{D}'$.

Consider the two functions

$$F_I(z_1,\dots,z_k) \equiv (\varphi, e^{z_1 Y_1} \dots e^{z_k Y_k} \psi) \tag{8.17}$$

$$F_{II}(z_1,\dots,z_k) \equiv (\varphi, e^{\sum_{i=1}^{k} f_i Y_i} e^{\sum_{i=1}^{\ell} g_i X_i} \psi) \tag{8.17'}$$

At first they are both defined for small real $z_1,\dots,z_k$ and are equal there.

F_I is analytic in each z_i in a vertical strip

$$S_\varepsilon^{(i)} \equiv \{z_i \in \mathbb{C} \mid |\mathrm{Re}\, z_i| < \varepsilon\}$$

if all z_r for $r \neq i$ are kept small and real.

Mapping these strips holomorphically onto horizontal strips

$$\hat{S}_\varepsilon^{(i)} \equiv \{\omega_i \in \mathbb{C} \mid |\mathrm{Im}\, \omega_i| < 2\pi\}$$

(by composing the maps tan and ar th with suitable scaling) we can use the Malgrange-Zerner-Stein-Kunze theorem [64] to obtain analyticity of F_I (and hence F_{II}) in a polydisc $\tilde{P}$ around the origin which we may identify with P .

For $(z_1,z_2,0,\dots,0) \in P$

$$F_I(z_1,z_2,0,\dots,0) = (e^{\bar{z}_1 Y_1}\varphi, e^{z_2 Y_2}\psi) \tag{8.18}$$

and for $z_1 = iy_1$, $z_2 = iy_2$ $(y_1,y_2 \in \mathbb{R})$

$$F_I(iy_1,iy_2,0,\dots,0) = (e^{-iy_1Y_1}\varphi, e^{iy_2Y_2}\psi) = (\varphi, e^{iy_1Y_1}e^{iy_2Y_2}\psi) \tag{8.19}$$

I claim that there are holomorphic curves in P obeying

$$\underline{f}(\underline{z}(t)) = e^t\underline{f}(\underline{z}(0)) \tag{8.20}$$

for $\operatorname{Re} t < 0$ and any real $z(0)$ (continuous up to the boundary).

Proof : By the Baker-Campbell-Hausdorff formula

$$\underline{f}(\underline{z}) = \underline{z} + O(|z|^2) .$$

So in a neighborhood of 0, $\underline{z} \to \underline{f}(\underline{z})$ may be inverted; our curves are then simply given by

$$\underline{z}(t) = \underline{f}^{-1}(e^t\underline{f}(\underline{z}(0)) \tag{8.21}$$

Along these curves we may analytically continue $F_I = F_{II}$, using just Theorem 8.3 and the fact that for $\varphi,\psi \in \mathcal{D}'$:

$$F_I(\underline{z}(t)) = (\exp(e^{\overline{t}}\sum_i f_i(z_i(0))Y_i)\varphi, e^{\sum_i g_iX_i}\psi) = F_{II}(\underline{z}(t)) . \tag{8.22}$$

Picking the curve determined by

$$\underline{z}\,(\tfrac{i\pi}{2}) = (iy_1,iy_2,0,\dots,0)$$

or equivalently

$$i\underline{f}(\underline{z}(0)) = \underline{f}(iy_1,iy_2,0,\dots,0)$$

we obtain

$$(\varphi, \exp(\sum_{j=1}^{k} f_j(iy_1,iy_2,0,\dots,0)Y_j) \quad \exp(\sum_{j=1}^{\ell} g_j(iy_1,iy_2,0,\dots,0)X_j)\psi) =$$

$$= (\varphi, e^{iy_1Y_1}e^{iy_2Y_2}\psi) \tag{8.23}$$

By density we may omit the vectors φ,ψ in (8.23) which then proves the crucial part of the multiplication law of G^* (we had to assume Y_1, Y_2 linearly independent, but if they are not everything is trivial anyhow).

To obtain the full group multiplication law one still has to show

$$e^X e^{iyY} e^{-X} = e^{iy\,\mathrm{ad}(e^X)(Y)} \tag{8.24}$$

for any $X \in h$, $Y \in m$. This follows easily by analytic continuation. We get the full miltiplication law of SO(d-1,1)

$$e^{iy_1Y_1} e^{X_1} e^{iy_2Y_2} e^{X_2} = e^{iY(y_1,y_2;Y_1,Y_2)} e^{X(y_1,y_2;Y_1,Y_2,X_1,X_2)} \tag{8.25}$$

(where Y(——) , X(——) are the functions given by the abstract multiplication law of G^* . (8.25) proves the representation property in a neighborhood U of $\mathbb{1}$. It is easy to see that it follows then for the whole group (by breaking up "large" group elements into suitable small ones in U). So we have proven :

Theorem 8.8 : On H there is a unitary representation of the proper orthochronous Lorentz group (or rather its universal covering group). It is uniquely determined by the representation of $U \subset E_d$ induced by the ordinary euclidean motions on $V_{\alpha_0,\varepsilon}$.

What is still lacking is the covariance of the translations. This is proven in a similar way; it is even somewhat easier. We omit the proof here (see [55]). One obtains

Theorem 8.9 : Denote by P_μ ($\mu = 0,\dots,d-1$) the generators of the translations; $P_0 = H$.

Then for any $\Lambda \in SO(d-1,1)$

$$U(\Lambda)^* e^{i\sum_\mu a_\mu P_\mu} U(\Lambda) = e^{i\sum_{\mu,\nu} a_\mu \Lambda_{\mu\nu} P_\nu} \tag{8.26}$$

From $P_0 \equiv H \geq 0$ and this covariance one obtains of course the spectral condition. The vacuum is given by the function $\mathbb{1}$ and is clearly invariant; its uniqueness is an obvious consequence of (S4) (clustering). So we checked all properties of Def. 8.1 and we have proven

Theorem 8.10 : From assumptions (S2), (S3), (S4) the existence of a relativistic Quantum Mechanics follows.

c) "Wightman Functions" and Their Analyticity.

To reveal the meaning of assumption (SO d) we look at the following operator $W^{(\varepsilon)}(C)$ on H : Let C be a loop in $\overline{\mathbb{R}}^d_+$ that touches the hyperplane $t = 0$ and has "temporal diameter" d . Let φ be given by the monomial

$$\prod_{i=1}^{n} W(C_i) \in V_+ \quad , \text{ i.e. } \quad \varphi = [\prod_{i=1}^{n} W(C_i)] \quad .$$

Then

$$W^{(\varepsilon)}(C)\varphi \equiv [W(C^{(\varepsilon,\vec{0})}) \prod_{i=1}^{n} W(C_i^{(d+2\varepsilon,\vec{0})})] \tag{8.27}$$

Maybe this is clearer pictorially :

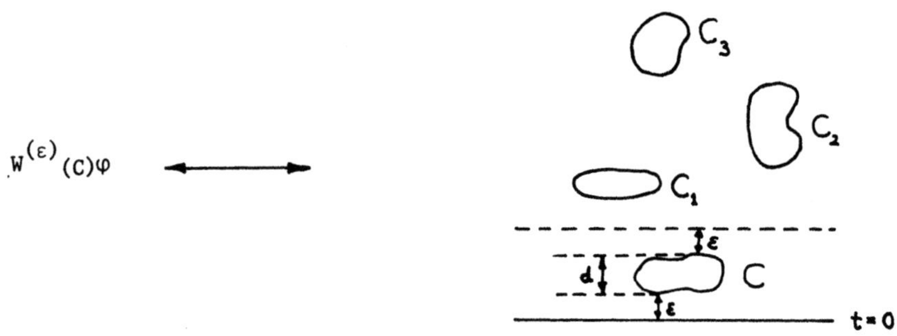

By linearity $W^{(\varepsilon)}(C)$ can be extended to V_+/N . We have now

<u>Lemma 8.11</u> : $\|W^{(\varepsilon)}(C)\| \leq K_{C,\varepsilon}$.

<u>Proof</u> : This follows easily by iterating Schwarz's inequality :

$$|(\varphi, W^{(\varepsilon)}(C)\psi)| \leq \|\varphi\| \ \|\psi\|^{1-2^{-2n}} \times (\psi, (W^{(\varepsilon)}(C)^* W^{(\varepsilon)}(C))^n \psi)^{\frac{1}{2n}}$$

where the lim sup of the last factor is bounded by $K_{C,\varepsilon}$ by (SO d) as $n \to \infty$.

□

(Strictly speaking one first has to prove this boundedness on V_+ which then permits to define $W^{(\varepsilon)}(C)$ on $\overline{V_+/N} = H$) .

Let now $C_1,\ldots,C_n$ be loops that are time-ordered and obey

$$d_t(C_{\ell-1},C_\ell) > 0 \quad , \quad \ell = 2,\ldots,n \ .$$

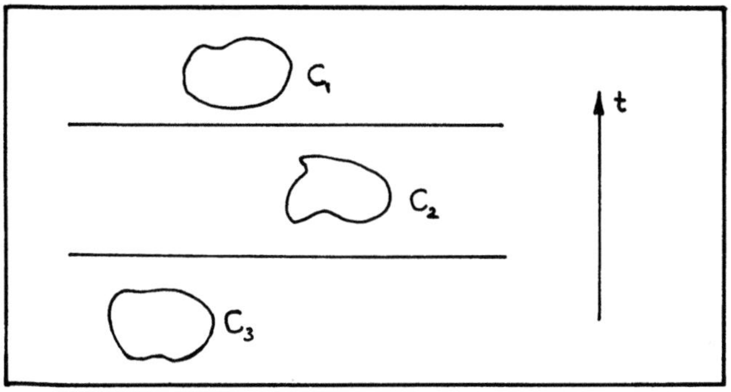

We define now a "Wightman function"

$$\mathcal{W}_{C_1,\ldots,C_n}(z_1,\ldots,z_n) \equiv S_n(C_n^{\tilde{z}_1},\ldots,C_n^{\tilde{z}_n}) \tag{8.28}$$

where

$$\tilde{z} = (iz_o,\vec{z}) \quad \text{if} \quad z = (z_o,\vec{z}) \ .$$

By the spectral condition $\mathcal{W}_{C_1,\ldots,C_n}(z_1,\ldots,z_n)$ is analytic in $z_1,\ldots,z_n$ as long as $\text{Im}(z_i-z_{i+1}) \in V_+$, $i = 1,2,3,\ldots,n-1$.

Note that $\mathcal{W}_{C_1,\ldots,C_n}$ does not have simple transformation properties under individual homogeneous Lorentz transformations of $C_1,\ldots,C_n$. Wilson loops contain all (integer) spins (half-integer ones come in when we study "stars" $B(C)$) . They should be considered as operators that create whole Regge trajectories, not just particles.

There is some joint analyticity of operator valued functions like

$$e^{i\vec{\alpha}\,\vec{B}}\, e^{ia_\mu P_\mu}\, W^{(\varepsilon)}(C)$$

but it would lead too far afield to analyze that here in detail.

But there is one more point that is important : $W^{(\varepsilon)}(C)$ can be boosted unitarily without harming the analyticity of the "Wightman functions" :

<u>Lemma 8.12</u> : Let C be a loop lying in the hyperplane $t = 0$. Then the following bilinear form $Q(\varphi,\psi)$ corresponds to a bounded operator of norm less or equal to

$K_{C,\varepsilon'}$, $(\varepsilon' = \varepsilon e^{-|\vec{\beta}|})$:

$$Q(\varphi,\psi) \equiv (e^{\varepsilon H} e^{-i\vec{\beta}\,\vec{B}} e^{-\varepsilon H} \varphi,\ W^{(\varepsilon)}(C) \quad e^{\varepsilon H} e^{-i\vec{\beta}\,\vec{B}} e^{-\varepsilon H} \psi) \tag{8.29}$$

which is defined on $H_{\varepsilon,o} \times H_{\varepsilon,o}$.

Proof : By the structure of $\tilde{P}_+^\uparrow$:

$$e^{\varepsilon H} e^{-i\vec{\beta}\,\vec{B}} e^{-\varepsilon H} = e^{\varepsilon H} (e^{-i\vec{\beta}\,\vec{B}} e^{-\varepsilon H} e^{i\vec{\beta}\,\vec{B}}) e^{-i\vec{\beta}\,\vec{B}} =$$

$$= e^{\varepsilon H} e^{-(\cosh|\vec{\beta}|)\varepsilon H + \frac{1}{|\vec{\beta}|}(\sinh|\vec{\beta}|)\varepsilon\vec{P}\cdot\vec{\beta}} \quad e^{-i\vec{\beta}\,\vec{B}} =$$

$$= e^{\varepsilon(1-e^{-|\vec{\beta}|})H} \exp\{-\varepsilon \sinh|\vec{\beta}|(H - \frac{\vec{\beta}}{|\vec{\beta}|} \cdot \vec{P})\}\ e^{-i\vec{\beta}\vec{\beta}}$$

The last two factors are bounded, so the lemma follows from the fact that

$$e^{\varepsilon(1-e^{-|\vec{\beta}|})H}\ W^{(\varepsilon)}(C)\ e^{\varepsilon(1-e^{-|\vec{\beta}|})H} = W^{(\varepsilon')}(C) \tag{8.30}$$

where $\varepsilon' = e^{-|\vec{\beta}|}\varepsilon$.

□

Remark : The bilinear form corresponds to an operator $W^{(\varepsilon)}(C_{(\vec{\beta})})$ where $C_{(\vec{\beta})}$ is the loop C boosted (in its own hyperplane) with boost parameter $\vec{\beta}$.

This means that we can define "Wightman functions" as before for loops $C_1,\dots,C_n$ that are lying in arbitrary space-like hyperplanes in Minkowski space.

d) Boundary Values ("Wightman Distributions")

This is now quite routine; we only have to remember to use a smoother test function space than S because of the possible strong exponential singularities that are allowed by (SO.c).

Theorem 8.13 : Let $C_1,\dots,C_n$ be loops lying in space-like hyperplanes. Then the boundary values

$$\lim_{\substack{\eta_1,\ldots,\eta_n \to 0 \\ \eta_i - \eta_{i+1} \in V_+ \\ i=1,\ldots,n-1}} \mathcal{W}_{C_1,\ldots,C_n}(x_1+i\eta_1,\ldots,x_n+i\eta_n)$$

exist in the sense of Jaffe ultradistributions (with indicator function $\omega(x)$ behaving like $|x|^{\frac{p}{p+1}}$ for $|x| \to \infty$, p as in (S0.c)) .

Remark : We do not want to go into the theory of Jaffe ultradistributions here, all that is needed is that they allow strict localization, i.e. for each open set there is a test function supported inside. See [65] for more details.

For a proof see [55].

e) Locality and Scattering Theory.

The following locality result is essentially a direct transfer of the proof by Jost [47] that symmetry of the Wightman functions implies locality. Some extra care is needed, however, due to the extended nature of the loops. See [55] for details.

Theorem 8.14 : Let C_1, C_2 be two loops, each lying in a spacelike hyperplane, and such that their convex hulls $\hat{C}_1$, $\hat{C}_2$ are spacelike to each other. Then the corresponding Wilson loops commute.

Remarks : (1) I suppose it is clear what is meant by the statement even though I didn't define all of the words in it properly.

(2) This locality result is not quite the best one could expect : it does not say, for instance, that the following pairs of loops in the $t = 0$ hyperplane commute, even though they should :

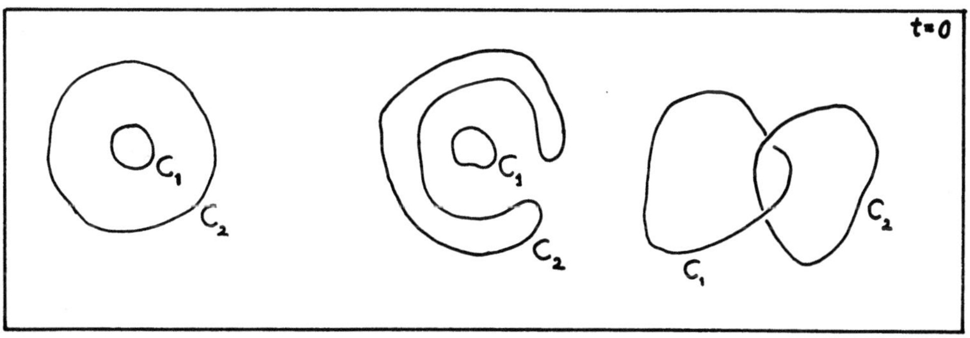

But the locality we proved is sufficient for scattering.

Idea of proof : By the assumption on $\hat{C}_1$, $\hat{C}_2$ all points in

$$D_{12} \equiv \{x_1 - x_2 | x_1 \in C_1 \, , \, x_2 \in C_2\}$$

are Jost points and there is a boost generator $\vec{\beta}\cdot\vec{B}$ such that

$$e^{\varepsilon\vec{\beta}\cdot\vec{B}} \begin{cases} \in V_+ & \text{for} \quad \varepsilon > 0 \\ \in V_- & \text{for} \quad \varepsilon < 0 \end{cases}$$

If we consider now a slightly ill-defined looking "Wightman function"

$$\mathcal{W}_{\tilde{C}_1,\ldots,\tilde{C}_k,C_1,C_2,\tilde{C}_{k+1},\ldots,\tilde{C}_n}(z_1,\ldots,z_k,0,0,z_{k+1},\ldots,z_n)$$

with $\quad \mathrm{Im}(z_i - z_{i+1}) \in V_+ \quad , \quad i = 1,\ldots,n-1$

and $\quad \mathrm{Im}\, z_k \in V_+ \quad , \quad \mathrm{Im}\, z_{k+1} \in V_-$

we see, using some arguments from complex analysis (see [55]), that this is really an analytic function by applying $e^{\vec{\beta}\cdot\vec{B}}$ to all arguments and loops. Doing this for $\varepsilon > 0$ we obtain C_1 before C_2 , for $\varepsilon < 0$, however, C_2 comes before C_1 . But by Lorentz invariance the function stays the same under the boost $e^{\varepsilon\vec{\beta}\cdot\vec{B}}$. Sending ε to zero and using continuity shows that we can interchange C_1 and C_2 . □

As remarked already, the main importance of this locality lies in the fact that it leads to strong decay of the truncated Wightman distributions in spacelike directions and thereby to strong asymptotic limits ("in" and "out" states) by the Haag-Ruelle theory, provided the energy-momentum spectrum contains an isolated one-particle hyperboloid ("upper mass gap"). Since this is well known (see [47,66,78] we just state

Theorem 8.15 : If there is an isolated one-particle hyperboloid in the spectrum of $(H, \vec{P})$ we obtain strong convergence to the asymptotic states and isometric Møller operators $\Omega^{\pm}$. The S-matrix $\Omega^{+}\,\Omega^{-*}$ is in general only a partial isometry. If we know that the theory is asymptotically complete, the S-operator is unitary.

This theorem is of course not too satisfactory but in this respect the situation is no worse than in ordinary quantum field theory. Methods to prove the upper mass gap are available [67,76] and can in principle be applied in this framework (maybe it would be useful to try them out on lattice gauge theories). I have nothing to say to the question of asymptotic completeness except that I hope the world does not resemble the "black flag roach motel" ("the roaches check in, but they don't check out").

This sketch is just intended to show that gauge invariant extended objects provide a framework for dynamics that looks essentially as reasonable as ordinary quantum field theory. And there is at least some hope that they might be a little easier to construct in $d = 4$ than quantum fields.

Appendix : The Geometric Setting of Gauge Theories.

The characteristic feature of gauge theories is the fact that at each point p of the space-time manifold M we have an internal symmetry space I. I is either a Lie group G, the gauge group (if we are dealing with the gauge field itself), or a vector space on which G is acting (if we are dealing with matter fields). So if we consider an open neighborhood U of a point $p \in M$, the space B on which the fields live looks like a direct product $U \times I$. This local direct product decomposition is called a local trivialization or in the language of physics a choice of a gauge. Different patches U, U' require a transition function $g_{U,U'}$ that relates the gauge choices in U and U' whenever $U \cap U' \neq \phi$: if $p \in U \cap U'$ we consider $(p,f) \in U \times I$ and $(p,f') \in U' \times I$ to be the same point in B provided f is related to f' by the action of $g_{U,U'}$. A patching (i.e. a covering by open sets) of M together with all transition functions defines B as a fiber bundle; I is called the fiber, M is called the base manifold. If $I = G$ we speak of a principal bundle, if I is a vector space we speak of an associated vector bundle. If M is not contractible, for instance if it has the topology of a sphere or a torus, B may turn out not to be homeomorphic to $M \times I$; in this case we say that B is nontrivial. In our context this is relevant for the case of the so-called (anti-) periodic boundary conditions corresponding to a torus as the base manifold M.

The concept of a gauge field corresponds geometrically to the concept of a connection in a principal bundle. A connection gives a prescription in which direction to move in the bundle B when a certain direction is given in the base space M, or put differently : A connection is a prescription to lift curves from M into B provided a starting point in B has been chosen. An equivalent description is the following: The tangent space of a point in B splits into a direct sum of the tangent space to the fiber I (the vertical directions) and its algebraic complement (the horizontal directions) which is isomorphic to the tangent space of a point in M ; a connection is just a (smooth) choice of horizontal subspace for each point. This horizontal subspace can be obtained as the null space of a Lie algebra valued 1-form ω, the connection form. In a local trivialization ω is determined by a Lie algebra valued 1- form $A = \Sigma A_\mu dx^\mu$ on M through the equation $\omega = g^{-1}Ag+g^{-1}dg$ $(g \in I = G$; more precisely $g^{-1}Ag+g^{-1}dg$ is the pullback of ω with respect to the local trivialization $\varphi : U \times I \to B$, $U \subset M$). The components A_μ are known as the Yang-Mills vector potential in the language of physics.

Normally a closed curve C in M will not lift to a closed curve in B. The endpoint will however lie on the same fiber as the starting point $\tilde{p}_o$ and thereby define an element $g_{\tilde{p}_o}(C)$ of G. $g_{\tilde{p}_o}(C)$ is called the holonomy operator corresponding to C and $\tilde{p}_o$. If a local trivialization (gauge) has been chosen such that $\tilde{p}_o = (p_o, \mathbf{1})$ the suggestive physicist's notation

$$g_{\tilde{p}_0}(C) = P \exp \int_C A$$

(where P stands for path ordering) can be used.

If $g_{\tilde{p}_0}(C)$ is not always the identity $\mathbb{1}$ we call the connection nontrivial; if this is the case for a contractible C we say the connection possesses curvature. The curvature form is a (horizontal) 2-form with values in the Lie algebra on B ; in a local trivialization it is determined by a Lie algebra valued 2-form F on M :

$$F = dA + \frac{1}{2}[A,A] .$$

The components of F are the Yang-Mills field strength tensor. If S(C) is a smooth surface bordered by C

$$\exp \int_{S(C)} F \cong g_{\tilde{p}_0}(C)$$

in leading order in $|S(C)|$. $g_{\tilde{p}_0}(C) = \mathbb{1} \in G$ for all C implies $F = 0$ and triviality of the principal bundle; on the other hand $F = 0$ implies $g_{\tilde{p}_0}(C) = \mathbb{1}$ for all contractible closed curves C .

Matter fields are given as sections of vector or spinor bundles. A section is simply a smooth assignment of a point $\tilde{p}$ in B to each $p \in M$ such that in a local trivialization $\tilde{p} = (p,f)$. So locally a section is just a function from M to I .

An important concept is the notion of topological charge density which is given by the mathematical concept of Chern classes. In a simple minded way the Chern classes c_n are defined by [I,90]

$$\det(1 + \frac{\lambda}{2\pi} F) = \Sigma \lambda^n c_n$$

so

$$c_1 = \frac{1}{2\pi} \operatorname{tr} F , \quad c_2 = \frac{1}{8\pi^2} \operatorname{tr} F \wedge F .$$

It can be shown that $\int_{M_{2n}} c_n$ is always an integer if M_{2n} is a compact 2n-dimensional submanifold of M without boundary (i.e. a 2n-cycle). $\int_{M_{2n}} c_n$ is called Chern number or topological charge.

The concepts of principal bundles, connection and curvature can easily be visualized by the following simple "classical" example : Consider a ball that is allowed to roll on a surface M ; assume that there are some patterns on the ball so its orientation can be observed. Its configuration space is locally just the direct product of its position space M and the space of orientations. The space of orientations may be taken as the space of orthonormal 3-frames fixed to the ball and can be identified with SO(3) . So the configuration space may be viewed as a principal bundle B with fiber SO(3) and base M .

A connection is now given by the "rolling" constraint : Moving the ball along a given curve will change its orientation in a definite way. This connection is non-trivial; the curvature F corresponds to infinitesimal rotations about the axis determined by the ball's center and the point of contact with M . For a flat surface and a "natural" choice of gauge

$$A = \frac{1}{R}\,(L_1 dx_2 - L_2 dx_1) \quad ; \quad F = -\frac{1}{R^2}\, L_3 dx_1 dx_2$$

where L_1, L_2 , L_3 are standard skew adjoint SO(3) generators and R is the ball's radius. We leave it to the reader to work out various holonomy operators in this example; unfortunately this example does not allow for non-vanishing Chern numbers.

REFERENCES

The following nonstandard abbreviations are used :

CMP = Communications in Mathematical Physics

JMP = Journal of Mathematical Physics

Brandeis 1965 = Axiomatic Field Theory, M. Chretien, S. Deser, eds., Gordon and Breach, New York 1966 (Brandeis lectures 1965).

Battelle 1971 = Statistical Mechanics and Mathematical Problems, Battelle Seattle 1971 Rencontres, A. Lenard, ed., Lecture Notes in Physics 20, Springer-Verlag, Berlin, Heidelberg, New York, 1973.

Erice 1973 = Constructive Quantum Field Theory, G. Velo and A.S. Wightman, eds. (Proceedings of the 1973 Erice Summer School), Lecture Notes in Physics 25, Springer-Verlag, Berlin, Heidelberg, New York 1973.

Erice 1975 = Renormalization Theory, G. Velo and A.S. Wightman, eds., D. Reidel, Dordrecht 1976 (Proceedings of the 1975 Erice School on Mathematical Physics).

Cargèse 1976 = New Developments in Quantum Field Theory and Statistical Mechanics, M. Lévy and P.K. Mitter, eds., Plenum Press, New York and London 1977 (Proceedings of the 1976 Cargèse Summer School).

Rome 1977 = Mathematical Problems in Theoretical Physics (Proceedings of the Rome $M \cap \phi$ conference 1977), G. dell'Antonio, S. Doplicher, G. Jona-Lasinio, eds., Lecture Notes in Physics 80, Springer-Verlag, Berlin, Heidelberg, New York 1978.

Schladming 1978 = Facts and Prospects of Gauge Theories (Proceedings of the 1978 Schladming Winter School), H. Mitter, ed., Acta Physica Austriaca, Suppl. XIX.

Esztergom 1979 = Colloquia Mathematica Societatis Janos Bolyai, 27. Random Fields (Esztergom, Hungary 1979), North Holland, Amsterdam 1981.

Kaiserslautern 1979 = Field Theoretical Methods in Particle Physics, W. Rühl, ed. (Proceedings of the 1979 Kaiserslautern Summer School), Plenum, New York 1980.

Lausanne 1979 = Mathematical Problems in Theoretical Physics, (Proceedings of the Lausanne $M \cap \phi$ conference 1979), K. Osterwalder, editor, Lecture Notes in Physics 116, Springer-Verlag, Berlin, Heidelberg, New York 1980.

Cargèse 1976 = New Developments in Quantum Field Theory and Statistical Mechanics, M. Lévy and P.K. Mitter, eds., Plenum Press, New York and London 1977 (Proceedings of the 1976 Cargèse Summer School).

Poiana Brasov 1981 = Gauge Theories : Fundamental Interactions and Rigorous Results (Proceedings of the 1981 Poiana Brasov Summer School), to appear at Birkhäuser, Boston.

REFERENCES TO PART I - "Lattice Gauge Theories"

[1] C. Becchi, A. Rouet and R. Stora, Ann. Phys. 98, 287 (1976) and Erice 1975.

[2] P. Breitenlohner and D. Maison, CMP 52, 11; 39; 55 (1977).

[3] T. Kugo and I. Oshima, Prog. Theor. Phys. 60, 1869 (1978); 61, 294 (1979); 61, 644 (1979).

[4] R. Jackiw, Rev. Mod. Phys. 52, 661 (1980); T. Eguchi, P.B. Gilkey, A.J. Hanson, Phys. Rep. 66, 213 (1980).

[5] F. Wegner, JMP 12, 2259 (1971).

[6] K. Wilson, Phys. Rev. D10, 2445 (1974).

[7] B. Julia and G. Toulouse, J. de Physique Lettres 16 (1979), L 395; G. Toulouse, Cargèse 1979.

[8] F. Strocchi, CMP 56, 57 (1979).

[9] J. Drouffe and C. Itzykson, Phys. Rep. 38C, 133 (1978).

[10] J. Kogut, Rev. Mod. Phys. 51, 659 (1979).

[11] B. Simon, The $P(\phi)_2$ (Euclidean) Quantum Field Theory, Princeton University Press, Princeton 1974.

[12] J. Glimm and A. Jaffe, Cargèse 1976; Quantum Physics, Springer, New York, Heidelberg, Berlin 1981.

[13] F. Guerra, L. Rosen and B. Simon, Ann. Math. 101, 111 (1975).

[14] Y.M. Park, JMP 16, 1065 (1975).

[15] E. Nelson, Topics in Dynamics I : Flows, Princeton University Press and University of Tokio Press, Princeton, N.J., 1969.

[16] J. Villain, J. de Physique 36, 581 (1975); A.M. Polyakov, Phys. Lett. B59, 79 (1975).

[17] J.M. Drouffe, Phys. Rev. D18, 1174 (1978).

[18] K. Wilson, Cargèse 1976.

[19] K. Osterwalder and E. Seiler, Ann. Phys. 110, 440 (1978); K. Osterwalder, Cargèse 1976; E. Seiler, Rome 1977.

[20] P. Hirschfeld, Nucl. Phys. B 157, 37 (1979).

[21] J. Fröhlich, B. Simon and T. Spencer, CMP 50, 79 (1976).

[22] J. Fröhlich and E. Lieb, CPM 60, 233 (1978).

[23] D.C. Brydges, J. Fröhlich and E. Seiler, Ann. Phys. 121, 227 (1979).

[24] M. Lüscher, CMP 54, 283 (1977).

[25] J. Fröhlich, R. Israel, E. Lieb and B. Simon, CMP 62, 1 (1978).

[26] G. 't Hooft, Nucl. Phys. B 153, 141 (1979); B 138, 1 (1978); Cargèse 1979.

[27] B. Simon, Phys. Rev. Lett. 36, 1083 (1976).

[28] J. Hess, R. Schrader and D. Uhlenbrock, Duke Math. Journ. 44, 893 (1977).

[29] B. Simon, Functional Integration and Quantum Physics, Academic Press, New York, 1979.

[30] W. Hunziker, Lausanne 1979.

[31] D.C. Brydges, J. Fröhlich and E. Seiler, Nucl. Phys. B 152, 521 (1979).

[32] I.O. Stamatescu, Max-Planck preprint MPI-PAE/PTH 60/80 , to appear in Phys.Rev.D.

[33] D. Hogreve, R. Schrader and R. Seiler, Nucl. Phys. B 142, 525 (1978).

[34] Y. Avron and B. Simon, Phys. Lett. 75A, 41 (1979).

[35] J. Ginibre, CMP 16, 310 (1970).

[36] G.F. de Angelis, D. de Falco, F. Guerra, CMP 57, 201 (1977), Lett. Nuovo Cim.18, 536 (1977); Phys. Lett. 68, 255 (1977); Phys. Rev. D17, 1624 (1978).

[37] C. Gruber, H. Kunz, CMP 22, 133; (1971).

[38] G. Gallavotti, F. Guerra, S. Miracle-Solé, Rome 1977.

[39] G. Gallavotti, A. Martin-Löf, S. Miracle-Solé, Battelle 1971.

[40] D. Ruelle, Statistical Mechanics, W.A. Benjamin, Inc., Reading, Massachusetts, 1969.

[41] R. Marra, S. Miracle-Solé, CMP 67, 233 (1979).

[42] S. Malyshev, CMP 64, 131 (1979).

[43] G.C. Rota, Z. Wahrsch. verw. Geb. 2, 340 (1964); O. Penrose in Statistical Mechanics, T. Bak, ed., Benjamin, New York, 1967.

[44] J.-P. Eckmann, J. Magnen, R. Sénéor, CMP 39, 251, (1975).

[45] J. Glimm, A. Jaffe, Nucl. Phys. B149, 49 (1979).

[46] G. Münster, Nucl. Phys. B180 [FS 2], 23 (1981); Phys. Lett. 95B, 59 (1980).

[47] G. Mack, CMP 65, 91 (1979).

[48] G. Mack, Cargèse 1979.

[49] E. Fradkin, S. Shenker, Phys. Rev. D19, 3682 (1979).

[50] J. Glimm, A. Jaffe, T. Spencer, Ann. Phys. 101, 610; 101, 631 (1976).

[51] H.B. Nielsen and M. Ninomiya, Nucl. Phys. B185, 20 (1981); B193, 173 (1981).

[52] K. Gawędzki, CMP 63, 31 (1978).

[53] J.L. Challifour, D.H. Weingarten, JMP, 19, 1134 (1978).

[54] M. Creutz, Phys. Rev. D21, 1006 (1980); D.J.E. Callawey, L.J. Carson, Phys. Rev. D25, 531 (1982); K.C. Bowler, G.S. Pawley, B.J. Pendleton, D.J. Wallace, G.W. Thomas, Phys. Lett. 104B, 481 (1981).

[55] H.A. Kramers, G. Wennier, Phys. Rev. 60, 252 (1941); H. Ceva, L. Kadanoff, Phys. Rev. B3, 3918 (1971).

[56] C. Gruber, A. Hintermann, D. Merlini : Group Analysis of Classical Lattice Systems, Lecture Notes in Physics 60, Springer Verlag, Berlin, Heidelberg, New York 1977.

[57] G. Gallavotti, A. Martin-Löf, S. Miracle-Solé, Battelle 1971.

[58] F.A. Berezin The Method of Second Quantization, Academic Press, New York 1966.

[59] J. M. Kosterlitz, D.J. Thouless, J. Physics C6, 1181 (1973).

[60] J. Fröhlich, T. Spencer, CMP 81, 527 (1981).

[61] S. Elitzur, R.B. Pearson, J. Shigemitsu, Phys. Rev. D19, 3698 (1979).

[62] R. Israel, C. Nappi, CMP 64, 177 (1979).

[63] F. Guerra, Phys. Rev. Lett. 28, 1213 (1972).

[64] A. Guth, Phys. Rev. D21, 2291 (1980).

[65] T. Banks, J. Kogut, R. Myerson, Nucl. Phys. B121, 493 (1977).

[66] B. Durhuus, J. Fröhlich, CMP 75, 103 (1980).

[67] G. Mack, V.B. Petkova, Ann. Phys. (N.Y.) 23, 442 (1979).

[68] J. Fröhlich, Phys. Lett. 83B, 195 (1979).

[69] A. Messager, S. Miracle-Solé, C. Pfister, CMP 58, 19 (1978).

[70] G. Mack, V.B. Petkova, Ann. Phys. (N.Y.) 125, 117 (1980).

[71] R.L. Dobrushin, S.B. Shlosman, CMP 42, 31 (1975).

[72] N.D. Mermin, J. Math. Phys. 8, 1061 (1967).

[73] N.D. Mermin, H. Wagner, Phys. Rev. Lett. 17, 1133 (1966).

[74] J. Fröhlich, B. Simon, S. Spencer, CMP 50, 79 (1976).

[75] C. Itzykson, M.E. Peskin, J.B. Zuber, Phys. Lett. 95B, 259 (1980); A., E. and P. Hasenfratz, Nucl. Phys. B180 [FS 2] , 353 (1981); J.M. Drouffe and J.B. Zuber, Nucl. Phys. B180 [FS 2], 253 and 264 (1980); M. Lüscher, G. Münster, P. Weisz, Nucl. Phys. B180 [FS 2], 1 (1981); G. Münster, P. Weisz, Nucl. Phys. B180 [FS 2], 13 (1981); M. Lüscher, Nucl. Phys. B180[FS 2], 317 (1981); J. Kogut, R.B. Pearson, J. Shigemitsu, Phys. Lett. 98B, 63 (1981).

[76] G. 't Hooft, Nucl. Phys. B153, 141 (1979); Cargèse (1979).

[77] J. Glimm, A. Jaffe, CMP 56, 195 (1977).

[78] J. Glimm, A. Jaffe, Phys. Lett. 66B, 67 (1977).

[79] J. Fröhlich, Y.M. Park, CMP 59, 131 (1977).

[80] Erice 1973.

[81] W. Kerler, Phys. Rev. D23, 2384 (1981).

[82] G. Parisi, Cargèse 1979; J.M. Drouffe, G. Parisi, N. Sourlas, Nucl. Phys. B161, 397 (1979).

[83] J.M. Drouffe, Nucl. Phys. B170 [FS1], 211 (1980).

[84] L. Susskind, Phys. Rev. D20, 2610 (1979).

[85] A.M. Polyakov, Phys. Lett. 72B, 477 (1978).

[86] G. Mack, Phys. Lett. 78B, 263 (1978).

[87] J. Fröhlich, T. Spencer, Massless Phases and Symmetry Restoration in Abelian Gauge Theories and Spin Systems, to appear in CMP (I.H.E.S. preprint 1981).

[88] S. Kobayashi, K. Nomizu, Foundations of Differential Geometry, Vols. I,II, Interscience, New York, 1963 and 1969.

[89] A. Trautman, Rep. Math. Phys. (Torun), 29 (1970).

[90] R. Bott, S.S. Chern, Acta Math. 115, 71 (1965).

[91] F. Guerra, Kaiserslautern 1979; G.F. de Angelis, D. de Falco, F. Guerra, R. Marra, Schladming 1978.

[92] E. Seiler, Poiana Brasov 1981.

[93] M. Göpfert, G. Mack, CMP 81, 97 (1981); 82, 545 (1982).

[94] G. Mack, H. Meyer, A Disorder Parameter that Tests for Confinement in Gauge Theories With Quark Fields, DESY preprint 81-043 (1981), to appear in Nucl. Phys. B.

[95] E. Seiler, I.O. Stamatescu, Lattice Fermions and θ-Vacua, to appear in Phys. Rev. D.

[96] M. Göpfert, Nucl. Phys. B190 [FS3], 151 (1981).

[97] N. Kawamoto, Nucl. Phys. B190 [FS3], 617 (1981).

[98] A.P. Wallace, Algebraic Topology, Homology and Cohomology, Benjamin, New York 1970.

REFERENCES TO PART II - "Continuum Gauge Quantum Field Theories"

[1] F. Nietzsche "Von den Hinterweltlern" in "Also sprach Zarathustra".

[2] R. Schrader, CMP 49, 131 (1976); 50, 97 (1976).

[3] C. Callan, Phys. Rev. D2, 1541 (1970), K. Symanzik, CMP 18, 227 (1970); Springer tracts in modern physics, vol. 57, 222 (1971).

[4] K. Osterwalder, R. Schrader, CMP 31, 83 (1973); 42, 281 (1975).

[5] E. Hewitt and K.A. Ross, Abstract harmonic analysis I, II, Springer-Verlag, New York 1970.

[6] H.G. Dosch, V.F. Müller, Fortschr. Phys. 27, 547 (1979).

[7] H.D. Politzer, Phys. Rev. Lett. 30, 1346 (1973); D. Gross, F. Wilczek, Phys. Rev. Lett. 30, 1343 (1973); see also D. Marciano, H. Pagels, Phys. Rep. 36, 137 (1978).

[8] M. Creutz, Phys. Rev. D21, 2308 (1979); Phys. Lett. 45, 313 (1980).

[9] T. Bałaban, Lausanne 1979; Ultraviolet Stability for a Model of Interacting Scalar and Vector Fields, Aarhus preprint No.19, 1981.

[10] G. Gallavotti, Mem. dell'Accad. dei Lincei XV, 23 (1978); G. Benfatto, M. Cassandro, G. Gallavotti, F. Nicolò, E. Olivieri, E. Presutti, E. Scacciatelli, CMP 59, 143 (1978); 71, 95 (1980).

[11] B. Simon, The $P(\phi)_2$ (Euclidean) Quantum Field Theory, Princeton University Press, Princeton 1974.

[12] J. Glimm, A. Jaffe, Cargèse 1976; Quantum Physics, Springer-Verlag, New York, Heidelberg, Berlin 1981.

[13] J. Fröhlich, E. Seiler, Helv. Phys. Acta 49, 889 (1976).

[14] M.T. Matthews, A. Salam, Nuovo Cim. 12, 367 (1948); Phys. Rev. 80, 440 (1950).

[15] E. Seiler, CMP 42, 163 (1975).

[16] J. Magnen, R. Sénéor, Lausanne 1979; work in preparation.

[17] J. Magnen, R. Sénéor, private communication; J. Magnen, Esztergom 1979.

[18] E. Seiler, B. Simon, Ann. Phys. (N.Y.) 97, 470 (1976).

[19] E. Seiler, B. Simon, CMP 45, 99 (1976); B. Simon, Trace Ideals and Their Applications, Cambridge University Press 1979.

[20] M. Reed, B. Simon, Methods of Modern Mathematical Physics IV, Academic Press, New York 1979; B. Simon, Adv. Math. 24, 244 (1977); N. Dunford, J.T. Schwartz, Linear Operators II, Interscience, New York and London 1963.

[21] J. Schwinger, Phys. Rev. 82, 664 (1951).

[22] H. Hogreve, R. Schrader, R. Seiler, Nucl. Phys. B142, 525 (1978).

[23] E. Seiler, Phys. Rev. D22, 2412 (1980).

[24] J. Schwinger, Phys. Rev. 128, 2425 (1962).

[25] W.I. Lenin, Materialism and Empiriocriticism. Actually he never said that. But he might have, had he lived long enough.

[26] J. Challifour, D. Weingarten, Ann. Phys. (N.Y.) 123, 61 (1979).

[27] D. Weingarten, Ann. Phys. 126, 154 (1980).

[28] D.C. Brydges, J. Fröhlich, E. Seiler, CMP 71, 159 (1980).

[29] ———————, CMP 79, 353 (1981).

[30] R. Schrader, CMP 58, 229 (1978).

[31] J. Potthoff, Thesis, F.U. Berlin 1980.

[32] D.C. Brydges, J. Fröhlich, E. Seiler, Ann. Phys. (N.Y.), 121, 227 (1979).

[33] M. Reed, B. Simon, Methods of Modern Mathematical Physics II, Academic Press, New York, 1973.

[34] E. Seiler, B. Simon JMP 16, 2289 (1975).

[35] J. Glimm, A. Jaffe, Fortschr. Phys. 21, 327 (1975).

[36] E. Nelson, Erice 1973.

[37] F. Guerra, L. Rosen, B. Simon, Adv. Math. 101, 111 (1975).

[38] J. Glimm, A. Jaffe, T. Spencer, Erice 1973.

[39] A. Cooper, L. Rosen, Trans. Amer. Math. Soc. 234, 93 (1977).

[40] J. Fröhlich, Helv. Phys. Acta 47, 265 (1974).

[41] J. Glimm, A. Jaffe, T. Spencer, CMP 45, 203 (1975).

[42] F. Guerra, Colloq. Int. CNRS 248, (1975).

[43] Y.M. Park, JMP 18, 354 (1977).

[44] J. Fröhlich, B. Simon, Ann. Math. 105, 493 (1977).

[45] F. Guerra, Phys. Rev. Lett. 28, 1213 (1972).

[46] J. Magnen, R. Sénéor, CMP 51, 297 (1976).

[47] R. Streater, A.S. Wightman, PCT, Spin & Statistics and All That, Benjamin/Cummings, Reading, Mass. 1964/1978; R. Jost, The General Theory of Quantized Fields, AMS, Providence, R.I. 1965.

[48] J. Fröhlich, Ann. Inst. H. Poincaré 21, 271 (1974); J. Glimm, A. Jaffe, Cargèse 1976.

[49] B. Plohr, Ph. D. thesis, Princeton 1980; JMP 22, 2184 (1981).

[50] A. Jaffe, C. Taubes, Vortices and Monopoles, Progress in Physics, vol. 2, Birkhäuser, Boston 1980.

[51] J.P. Eckmann, J. Magnen, R. Sénéor, CMP 39, 251 (1975).

[52] T. Bałaban, K. Gawedzki, A low temperature expansion for the pseudoscalar Yukawa model of quantum fields in two space-time dimensions, to appear in Ann. Inst. H. Poincaré (Harvard preprint 1980).

[53] J. Schwinger, Phys. Rev. 82, 664 (1951), S.L. Adler, Phys. Rev. 177, 2426 (1967), S.L. Adler, W. Bardeen, Phys. Rev. 182, 1515 (1969).

[54] J.-L. Gervais, A. Neveu, Phys. Lett. 80B, 255 (1979); Cargèse 1979; Nucl. Phys. B153, 445 (1979); B163, 189 (1980); Y. Nambu, Phys. Lett. 80B, 372 (1979), A.M. Polyakov, Phys. Lett. 82B, 247 (1979); Nucl. Phys. B164, 171 (1979), Yu.M. Makeenko, A.A. Migdal, Phys. Lett. 88B, 135 (1979).

[55] J. Fröhlich, K. Osterwalder, E. Seiler, in preparation; J. Fröhlich, Cargèse 1979.

[56] M. Kaku, K. Kikkawa, Phys. Rev. D10, 1110 (1974).

[57] E. Cremmer, J.-L. Gervais, Nucl. Phys. B76, 209 (1974).

[58] A. Chodos, R.L. Jaffe, K. Johnson, C.B. Thorn, V.F. Weisskopf, Phys. Rev. D9, 3471 (1974).

[59] J. Fröhlich, Adv. Appl. Math. 1, 237 (1980); A. Klein, L.J. Landau, J. Funct. Anal. 44, 121 (1981).

[60] V. Glaser, in Problems in Theoretical Physics, p. 69, Moscow, Nauka 1969.

[61] E. Nelson, Ann. Math. 81, 547 (1959).

[62] J. Simon, CMP 28, 39 (1972); M. Flato, J. Simon, H. Snellman, D. Sternheimer, Ann. Sci. ENS (Paris) 5, 3 (1972).

[63] S. Helgason, Differential Geometry, Lie groups and Symmetric spaces, Academic Press, New York 1978; S. Kobayashi and K. Nomizu, Foundations of Differential Geometry, Interscience, New York 1963-69.

[64] H. Epstein, Brandeis 1965; E. Stein, H. Kunze, Amer. J. Math. 83, 723 (1960) (Section 8).

[65] A. Jaffe, Phys. Rev. 158, 1454 (1967); F. Constantinescu, W. Thalheimer, CMP 38, 299 (1974).

[66] K. Hepp, Brandeis 1965.

[67] T. Spencer, CMP 44, 143 (1975); J. Glimm, A. Jaffe, T. Spencer, Ann. Math. 100, 585 (1974); J.Z. Imbrie, CMP 78, 169 (1980); H. Koch, Ann. Inst. H. Poincaré 31, 173 (1979).

[68] H. Euler, W. Heisenberg, Z. Phys. 98, 714 (1936).

[69] Erice 1973.

[70] W. Kerler, Phys. Rev. D23, 2384 (1981).

[71] E. Seiler, I.O. Stamatescu, Lattice Fermions and θ-Vacua, to appear in Phys. Rev. D.

[72] M. Göpfert, G. Mack, CMP 82, 545 (1982).

[73] A. Patrascioiu, E. Seiler, I.O. Stamatescu, Phys. Lett. 107B, 364 (1981).

[74] E. Seiler, Poiana Brasov 1981.

[75] K.R. Itô, Construction of Euclidian QED_2 via Lattice Gauge Theory, Bedford College preprint 1981, to appear in CMP.

[76] R.S. Schor, CMP 59, 213 (1978).

[77] H.S. Sharatchandra, Phys. Rev. D18, 2042 (1978).

[78] M. Reed, B. Simon, Methods of Modern Mathematical Physics III, Academic Press, New York 1978.

[79] A.M. Garsia in : Proceedings of the 6th Berkeley Symposium on Mathematical Statistics and Probability, 2, L. Le Cam, J. Neyman, E.L. Scott, eds., Univ. of California Press, Berkeley, Cal. 1971.

Lecture Notes in Physics

Vol. 136: The Role of Coherent Structures in Modelling Turbulence and Mixing. Proceedings 1980. Edited by J. Jimenez. XIII, 393 pages. 1981.

Vol. 137: From Collective States to Quarks in Nuclei. Edited by H. Arenhövel and A. M. Saruis. VII, 414 pages. 1981.

Vol. 138: The Many-Body Problem. Proceedings 1980. Edited by R. Guardiola and J. Ros. V, 374 pages. 1981.

Vol. 139: H. D. Doebner, Differential Geometric Methods in Mathematical Physics. Proceedings 1981. VII, 329 pages. 1981.

Vol. 140: P. Kramer, M. Saraceno, Geometry of the Time-Dependent Variational Principle in Quantum Mechanics. IV, 98 pages. 1981.

Vol. 141: Seventh International Conference on Numerical Methods in Fluid Dynamics. Proceedings. Edited by W. C. Reynolds and R. W. MacCormack. VIII, 485 pages. 1981.

Vol. 142: Recent Progress in Many-Body Theories. Proceedings. Edited by J. G. Zabolitzky, M. de Llano, M. Fortes and J. W. Clark. VIII, 479 pages. 1981.

Vol. 143: Present Status and Aims of Quantum Electrodynamics. Proceedings, 1980. Edited by G. Gräff, E. Klempt and G. Werth. VI, 302 pages. 1981.

Vol. 144: Topics in Nuclear Physics I. A Comprehensive Review of Recent Developments. Edited by T.T.S. Kuo and S.S.M. Wong. XX, 567 pages. 1981.

Vol. 145: Topics in Nuclear Physics II. A Comprehensive Review of Recent Developments. Proceedings 1980/81. Edited by T. T. S. Kuo and S. S. M. Wong. VIII, 571-1.082 pages. 1981.

Vol. 146: B. J. West, On the Simpler Aspects of Nonlinear Fluctuating. Deep Gravity Waves. VI, 341 pages. 1981.

Vol. 147: J. Messer, Temperature Dependent Thomas-Fermi Theory. IX, 131 pages. 1981.

Vol. 148: Advances in Fluid Mechanics. Proceedings, 1980. Edited by E. Krause. VII, 361 pages. 1981.

Vol. 149: Disordered Systems and Localization. Proceedings, 1981. Edited by C. Castellani, C. Castro, and L. Peliti. XII, 308 pages. 1981.

Vol. 150: N. Straumann, Allgemeine Relativitätstheorie und relativistische Astrophysik. VII, 418 Seiten. 1981.

Vol. 151: Integrable Quantum Field Theory. Proceedings, 1981. Edited by J. Hietarinta and C. Montonen. V, 251 pages. 1982.

Vol. 152: Physics of Narrow Gap Semiconductors. Proceedings, 1981. Edited by E. Gornik, H. Heinrich and L. Palmetshofer. XIII, 485 pages. 1982.

Vol. 153: Mathematical Problems in Theoretical Physics. Proceedings, 1981. Edited by R. Schrader, R. Seiler, and D.A. Uhlenbrock. XII, 429 pages. 1982.

Vol. 154: Macroscopic Properties of Disordered Media. Proceedings, 1981. Edited by R. Burridge, S. Childress, and G. Papanicolaou. VII, 307 pages. 1982.

Vol. 155: Quantum Optics. Proceedings, 1981. Edited by C.A. Engelbrecht. VIII, 329 pages. 1982.

Vol. 156: Resonances in Heavy Ion Reactions. Proceedings, 1981. Edited by K.A. Eberhard. XII, 448 pages. 1982.

Vol. 157: P. Niyogi, Integral Equation Method in Transonic Flow. XI, 189 pages. 1982.

Vol. 158: Dynamics of Nuclear Fission and Related Collective Phenomena. Proceedings, 1981. Edited by P. David, T. Mayer-Kuckuk, and A. van der Woude. X, 462 pages. 1982.

Vol. 159: E. Seiler, Gauge Theories as a Problem of Constructive Quantum Field Theory and Statistical Mechanics. V, 192 pages. 1982.

Selected Issues from

Lecture Notes in Mathematics

Vol. 728: Non-Commutative Harmonic Analysis. Proceedings, 1978. Edited by J. Carmona and M. Vergne. V, 244 pages. 1979.

Vol. 729: Ergodic Theory. Proceedings 1978. Edited by M. Denker and K. Jacobs. XII, 209 pages. 1979.

Vol. 730: Functional Differential Equations and Approximation of Fixed Points. Proceedings, 1978. Edited by H.-O. Peitgen and H.-O. Walther. XV, 503 pages. 1979.

Vol. 731: Y. Nakagami and M. Takesaki, Duality for Crossed Products of von Neumann Algebras. IX, 139 pages. 1979.

Vol. 733: F. Bloom, Modern Differential Geometric Techniques in the Theory of Continuous Distributions of Dislocations. XII, 206 pages. 1979.

Vol. 735: B. Aupetit, Propriétés Spectrales des Algèbres de Banach. XII, 192 pages. 1979.

Vol. 738: P. E. Conner, Differentiable Periodic Maps. 2nd edition, IV, 181 pages. 1979.

Vol. 742: K. Clancey, Seminormal Operators. VII, 125 pages. 1979.

Vol. 755: Global Analysis. Proceedings, 1978. Edited by M. Grmela and J. E. Marsden. VII, 377 pages. 1979.

Vol. 756: H. O. Cordes, Elliptic Pseudo-Differential Operators – An Abstract Theory. IX, 331 pages. 1979.

Vol. 760: H.-O. Georgii, Canonical Gibbs Measures. VIII, 190 pages. 1979.

Vol. 762: D. H. Sattinger, Group Theoretic Methods in Bifurcation Theory. V, 241 pages. 1979.

Vol. 765: Padé Approximation and its Applications. Proceedings, 1979. Edited by L. Wuytack. VI, 392 pages. 1979.

Vol. 766: T. tom Dieck, Transformation Groups and Representation Theory. VIII, 309 pages. 1979.

Vol. 771: Approximation Methods for Navier-Stokes Problems. Proceedings, 1979. Edited by R. Rautmann. XVI, 581 pages. 1980.

Vol. 773: Numerical Analysis. Proceedings, 1979. Edited by G. A. Watson. X, 184 pages. 1980.

Vol. 775: Geometric Methods in Mathematical Physics. Proceedings, 1979. Edited by G. Kaiser and J. E. Marsden. VII, 257 pages. 1980.

Vol. 779: Euclidean Harmonic Analysis. Proceedings, 1979. Edited by J. J. Benedetto. III, 177 pages. 1980.

Vol. 780: L. Schwartz, Semi-Martingales sur des Variétés, et Martingales Conformes sur des Variétés Analytiques Complexes. XV, 132 pages. 1980.

Vol. 782: Bifurcation and Nonlinear Eigenvalue Problems. Proceedings, 1978. Edited by C. Bardos, J. M. Lasry and M. Schatzman. VIII, 296 pages. 1980.

Vol. 783: A. Dinghas, Wertverteilung meromorpher Funktionen in ein- und mehrfach zusammenhängenden Gebieten. Edited by R. Nevanlinna and C. Andreian Cazacu. XIII, 145 pages. 1980.

Vol. 786: I. J. Maddox, Infinite Matrices of Operators. V, 122 pages. 1980.

Vol. 787: Potential Theory, Copenhagen 1979. Proceedings, 1979. Edited by C. Berg, G. Forst and B. Fuglede. VIII, 319 pages. 1980.

Vol. 791: K. W. Bauer and S. Ruscheweyh, Differential Operators for Partial Differential Equations and Function Theoretic Applications. V, 258 pages. 1980.

Vol. 792: Geometry and Differential Geometry. Proceedings, 1979. Edited by R. Artzy and I. Vaisman. VI, 443 pages. 1980.

Vol. 793: J. Renault, A Groupoid Approach to C*-Algebras. III, 160 pages. 1980.

Vol. 798: Analytic Functions, Kozubnik 1979. Proceedings. Edited by J. Ławrynowicz. X, 476 pages. 1980.

Vol. 799: Functional Differential Equations and Bifurcation. Proceedings 1979. Edited by A. F. Izé. XXII, 409 pages. 1980.

Vol. 801: K. Floret, Weakly Compact Sets. VII, 123 pages. 1980.

Vol. 802: J. Bair, R. Fourneau, Etude Géometrique des Espaces Vectoriels II. VII, 283 pages. 1980.

Vol. 804: M. Matsuda, First Order Algebraic Differential Equations. VII, 111 pages. 1980.

Vol. 805: O. Kowalski, Generalized Symmetric Spaces. XII, 187 pages. 1980.

Vol. 807: Fonctions de Plusieurs Variables Complexes IV. Proceedings, 1979. Edited by F. Norguet. IX, 198 pages. 1980.

Vol. 809: I. Gumowski and Ch. Mira, Recurrences and Discrete Dynamic Systems. VI, 272 pages. 1980.

Vol. 810: Geometrical Approaches to Differential Equations. Proceedings 1979. Edited by R. Martini. VII, 339 pages. 1980.

Vol. 816: L. Stoica, Local Operators and Markov Processes. VIII, 104 pages. 1980.

Vol. 819: Global Theory of Dynamical Systems. Proceedings, 1979. Edited by Z. Nitecki and C. Robinson. IX, 499 pages. 1980.

Vol. 821: Statistique non Paramétrique Asymptotique. Proceedings, 1979. Edited by J.-P. Raoult. VII, 175 pages. 1980.

Vol. 826: Ph. G. Ciarlet et P. Rabier, Les Equations de von Kármán. VI, 181 pages. 1980.

Vol. 827: Ordinary and Partial Differential Equations. Proceedings, 1978. Edited by W. N. Everitt. XVI, 271 pages. 1980.

Vol. 831: Representation Theory I. Proceedings, 1979. Edited by V. Dlab and P. Gabriel. XIV, 373 pages. 1980.

Vol. 832: Representation Theory II. Proceedings, 1979. Edited by V. Dlab and P. Gabriel. XIV, 673 pages. 1980.

Vol. 836: Differential Geometrical Methods in Mathematical Physics. Proceedings, 1979. Edited by P. L. García, A. Pérez-Rendón, and J. M. Souriau. XII, 538 pages. 1980.

Vol. 837: J. Meixner, F. W. Schäfke and G. Wolf, Mathieu Functions and Spheroidal Functions and their Mathematical Foundations Further Studies. VII, 126 pages. 1980.

Vol. 838: Global Differential Geometry and Global Analysis. Proceedings 1979. Edited by D. Ferus et al. XI, 299 pages. 1981.

Vol. 840: D. Henry, Geometric Theory of Semilinear Parabolic Equations. IV, 348 pages. 1981.

Vol. 841: A. Haraux, Nonlinear Evolution Equations- Global Behaviour of Solutions. XII, 313 pages. 1981.

Vol. 842: Séminaire Bourbaki vol. 1979/80. Exposés 543–560. IV, 317 pages. 1981.

Vol. 843: Functional Ánalysis, Holomorphy, and Approximation Theory. Proceedings. Edited by S. Machado. VI, 636 pages. 1981.

Vol. 845: A. Tannenbaum, Invariance and System Theory: Algebraic and Geometric Aspects. X, 161 pages. 1981.

Vol. 849: P. Major, Multiple Wiener-Itô Integrals. VII, 127 pages. 1981.

GPSR Compliance

The European Union's (EU) General Product Safety Regulation (GPSR) is a set of rules that requires consumer products to be safe and our obligations to ensure this.

If you have any concerns about our products, you can contact us on ProductSafety@springernature.com

In case Publisher is established outside the EU, the EU authorized representative is:

Springer Nature Customer Service Center GmbH
Europaplatz 3
69115 Heidelberg, Germany

Batch number: 10683525

Printed by Printforce, the Netherlands